Organometallic Compounds in the Marine Environment

Tarek Othman Said • Gehan Mohamed El Zokm

Organometallic Compounds in the Marine Environment

Analytical and Ecological Aspects

Springer

Tarek Othman Said
Marine Environment
National Institute of Oceanography and Fisheries
Cairo, Egypt

Gehan Mohamed El Zokm
Marine Environment
National Institute of Oceanography and Fisheries
Cairo, Egypt

ISBN 978-3-031-99188-2 ISBN 978-3-031-99189-9 (eBook)
https://doi.org/10.1007/978-3-031-99189-9

This Springer imprint is published by the registered company Springer Nature Switzerland AG
The registered company address is: Gewerbestrasse 11, 6330 Cham, Switzerland

Preface

An organometallic compound (OMC) is a transitional form between traditional inorganic and organic derivatives with a unique M-C bond. The book explores the occurrence, chemical structure, and a brief survey of OMCs in aquatic systems, including water, sediment, and biota. An essential goal of the book is to thoroughly discuss the effects of OMCs on the marine ecosystem and their sources—extraction methodologies regarding stability, recoveries, and improving the reliability of speciation analysis. Analytical techniques involving the coupling of a separation technique such as gas chromatography, high-performance liquid chromatography, or supercritical fluid chromatography with a precise detector such as an emission, mass, or atomic absorption spectrometer are illustrated. The cold-vapor atomic fluorescence for Hg determination will be highlighted. The most recent approaches and strategies created for evaluating the level of OMCs in marine environments are discussed.

The evaluation of OMCs' environmental implications in light of the principles of "green chemistry" is a major focus of this work. Mechanisms governing OMCs' chemical behavior in the marine environment are addressed. The ecotoxicology of OMCs is a point of concern, focusing on modes of action. Ecotoxicological aspects of OMCs are deeply discussed with special reference to ecological risk assessment (RQ = PEC/PNEC) to create early warning systems. Influence of environmental factors (e.g., pH and salinity) and adsorbents in the matrices (e.g., minerals, organic carbons, and quartz) on the chemical behavior and toxicity of OMCs is addressed. Ca^{2+}-dependent and Mg-ATPase activity are two of the most acceptable mechanisms describing the toxicity of organotin compounds. Human health risk via tolerable average residue level (TARL), toxic unit (TU), cancer risk (CR), risk quotient (RQ), screening level (SL), target hazard quotient (THQ), and probabilistic distribution ($Intake_{i,k}$ (x, y)) are evaluated. Biosensing potential of OMCs in aquatic systems is assessed, and two examples of biomarkers of tributyltin (TBT) in marine invertebrates (Molecular Genetic and Imposex) are discussed. An overview highlighting the cutoff points of the ecological evaluation technique for OMCs is presented. Molecular modeling studies for describing ecotoxicity will be discussed. Our book highlights the fate and behavior of OMC transportation in various environmental compartments. The affinity of OMCs in different trophic states is assessed. The worldwide distribution of OMCs with a focus on hotspots in aquatic systems is displayed. The target groups are postgraduate students, academics, industry scientists, ecologists, environmentalists, and audiences from different disciplines interested in marine pollutants and OMCs.

Research on OMCs will take a multidisciplinary approach in the future, focusing on sustainability, innovation, and the creation of new applications that significantly affect society, technology, and science. Socioeconomic research on the uses of OMCs will benefit society and decision-makers alike. This book discusses the topic either as a stand-alone review or in the topic area's sequential order. Finally, this book covers the issue and will motivate the audience to address these hazardous materials. As a comprehensive study, there is a dearth of reliable quantitative information regarding the ecological and human toxicity of organometals (loids).

This book presents an overview of the recent progress made in the field of organometallic compounds, specifically focusing on their potential uses as sensors in aquatic ecosystems. The book highlighted the control (prevention) of OMCs. Added to the above, the book sheds light on achieving Sustainable Development Goals (SDGs), clean water and sanitation, and life below water. Additionally, we will assess joint efforts by different networks and worldwide policymakers to draw attention to the significance of this hot topic. The information presented in the book would be of particular interest to researchers in the marine field.

Cairo, Egypt — Prof Tarek Othman Said
Cairo, Egypt — Prof Gehan Mohamed El Zokm
2025

Competing Interests The authors have no competing interests to declare that are relevant to the content of this manuscript.

Contents

About the Authors

Prof Tarek Othman Said is a Professor in Marine and Analytical Chemistry, NIOF, Mediterranean Branch, Alexandria-Egypt. He got his B.Sc. is an excellent degree (1988), M.Sc. in Chemistry (July 1992), and Ph.D. in Analytical Chemistry (January 1997) from the Department of Chemistry-Faculty of Science, Mansoura University, Egypt. He has more than 35 years of experience in both environmental and analytical chemistry. He has played an essential role in the establishment of the NIOF central laboratories, the Mediterranean branch's infrastructure development, and El-Max Applied Research Station (NIOF-Mediterranean Sea Branch). In addition to creating the curriculum for over a decade at the Chemistry Department, Faculty of Science, KKU, Abha-KSA. He has experience in teaching different chemistry courses for undergraduate and postgraduate students in the faculties of science and engineering of different universities in Egypt and KSA and supervised different Ph.D. and M.Sc. programs. He has created a new Ph.D. program in the Chemistry Department, coordinated the M.Sc. program, chaired scientific research in the Chemistry Department, and served as a member of the College of Science, supervising numerous M.Sc. and Ph.D. students in Saudi and Egyptian universities. He is a member of the Editorial Board of Development in Analytical Chemistry, and an Associate Editor of *Frontiers in Environmental Science—Water and Wastewater Management*. He is an editorial board member of the *Egyptian Journal of*

Aquatic Research, EJAR and Editorial Manager of *Blue Economy Journal*. His research interests are pesticides, polycyclic aromatic hydrocarbons, polychlorinated biphenyls, and heavy metals in the Mediterranean Sea and Red Sea (Egypt), investigated and correlated with pollution and environmental aspects. Organo-tin and organo-lead in the Red Sea and the Mediterranean Sea were investigated and discussed through M.Sc. and Ph.D. theses. Different oil dispersants were tested by him through bioassay studies to calculate their efficiency. Bio-remedial treatment of organic pollutants using bacteria. He has performed speciation of heavy metals in different environmental samples; analysis of environmental samples for hydrocarbons, pesticides, and PCBs; and speciation of heavy metals in marine sediment, dust, soil, and biota samples, using hyphenated techniques (GC, GC/MS, GC/MS/MS, HPLC) for analysis of organic pollutants; AAS, ICP-OES; and ICP-MS for heavy metal and organometallic compounds analysis, and UV/V spectrophotometer for the analysis of inorganic pollutants. He is Supervisor of the central lab at NIOF, Alexandria-Egypt. He has demonstrated success in management roles, with a strong emphasis on relationship building and elemental analysis. He is adept at providing risk assessment and human health, overseeing environmental operations, and writing scientific reports. About 100 manuscripts in environmental and analytical chemistry have been published and acknowledged for surpassing risks to health and the environment and using alternative wastewater treatment methods. He is dedicated to lifelong learning and career advancement. He is eager to contribute to forward-thinking research in the field of health risks and environmental affairs. He is sharing in organizing different workshops, conferences, and international meetings belonging to NIOF, Egypt, and KKU, KSA. His Researcher Score = 26.07, H-Index = 27, i10-index = 51, and citations = 1907.

Prof Gehan Mohamed El Zokm is a Professor at the Marine Chemistry Department, Head of the Marine Environment Division, and Member of the Board of Directors of the National Institute of Oceanography and Fisheries (NIOF). She is the leader of the halogen, sulfide modeling, and surfactant groups at the NIOF, Mediterranean Sea Branch in Egypt. She got her B.Sc., M.Sc., and Ph.D. from the Chemistry Department at the Faculty of Science of Alexandria University, Egypt, both of which concerned heavy metal chemistry (focusing on preconcentration and fractionation tools). Her research interests are in analytical, instrumental, and environmental chemistry. Statistical analyses were the most prominent features of her research, which was conducted using the most recent software packages. Her papers discuss the topics of halogen chemistry, water quality, sulfur-sulfide chemistry, eutrophication, and the responses of marine biota to pollutants. Her researches also focus on the mobility and fluxes of various parameters within the marine environment. Extraction of natural products with many applications and nutritional studies is her recent project. Her research is also focused on water treatment and monitoring of pollutants (heavy metals, nutrients, phenols, hydrocarbons, and pesticides). She applied factorial design to water treatment experiments. She is interested in the ecotoxicity of surfactants on Egyptian coasts. She reviews many papers for important publishers (Elsevier, Springer, and Taylor and Francis). She headed and participated in several projects related to the marine environment. She supervised many master's and Ph.D. students, participated in conferences and workshops, and had several activities. She has provided a range of environmental consultations in the field of the marine environment. She heads the Master's and Doctoral Thesis Registration Committee at the National Institute of Oceanography and Fisheries. She has 70 research papers and four books with an h-index of 20.

Abbreviations

AAS	Atomic absorption spectrometry
ADI	Acceptable daily intake
AES	Atomic emission spectrometry
AFS	Atomic fluorescence spectrometry
AFSC	Anti-fouling systems on ships convention
AGC	Analytical green chemistry
APCs	Alkyl lead compounds
ATCs	Alkyl tin compounds
BTC	Butyl tin compounds
CLAW	Control of lead at work
CMMS	Computerized maintenance management systems
CRM	Certified reference material
CV	Cold vapor
CV-AAS	Cold vapor-atomic absorption spectrometry
DCP	Direct current plasma
DCP-AES	Direct current plasma-atomic emission spectrometry
DDTs	Dichlorodiphenyltrichloroethane
DIN	Direct injection nebulizer
DMA	Dimethyl arsenic acid
DMD	Dimethyl dioxirane
DMM	Dimethyl mercury
DMT	Dimethyl tin
DOC	Dissolved organic carbon
DOM	Dissolved organic matter
DPT	Diphenyl tin
EAN	Effective atomic number
ECD	Electron capture detection
EDI	Estimated daily intake
EPA	Environmental Protection Agency
EQO	Environmental quality objectives

Et_3Pb	Triethyl lead
FPD	Flame photometric detector
GAAS	Graphite atomic absorption spectrophotometer
GBMS	Global biotic mercury synthesis
GC	Gas chromatography
GC-AAS	Gas chromatography-atomic absorption spectrometry
GC-ECD	Gas chromatography-electron capture detector
GC-FID	Gas chromatograph-flame ionization detector
GC-MIP-AES	Gas chromatography-microwave induced plasma-AES
GF-AAS	Graphite furnace-atomic absorption spectrophotometer
GRs	Grignard reagents
HCHs	Hexachlorocyclohexanes
Hg	Mercury
HG-AAS	Hydride generation-AAS
HPLC	High-performance liquid chromatography
HQ	Hazard quotient
HSAB	Hard and soft acids and bases
HSE	Health and Safety Executive
iAs	Inorganic arsenic
ICP	Inductively coupled plasma
ICP-MS	Inductively coupled plasma-mass spectrometry
ICP-OES	Inductively coupled plasma-optical emission spectrometry
IMO	International Maritime Organization
IUPAC	International Union of Pure and Applied Chemistry
JECFA	Joint FAO/WHO Expert committee on food additives
MARPOL	International Convention for the Prevention of Pollution from Ships
MBT	Monobutyl tin
M-C	Metal-carbon
Me_3Pb	Trimethyl lead
Me_4Pb	Tetramethyl lead
MeHg	Methyl mercury
MEM	Mono-ethyl mercury
MIP	Microwave-induced plasma
MIP-AES	Microwave-induced plasma-atomic emission spectrometry
MIP-OES	Microwave-induced plasma-optical emission spectrometry
MMA	Monomethyl arsenic acid
MMM	Mono-methyl mercury
MMT	Methylcyclopentadienyl manganese tricarbonyl
MPM	Mono-phenyl mercury
MS	Mass spectrometry
MSFD	Marine Strategy Framework Directive
MTCs	Methyl tin compounds
$NaB(Ph)_4$	Sodium tetraphenyl lead borate
$NaBEt_4$	Sodium tetraethyl borate
NOAA	National Oceanic and Atmospheric Administration

NORMAN	Network of reference laboratories, research centers, and related organizations for monitoring of emerging environmental substances
OHgCs	Organomercury compounds
OLC	Organolead compounds
OLiC	Organolithium compounds
OMCs	Organometallic compounds
OMgC	Organomagnesium compounds
OPbCs	Organolead compounds
OSPAR	Convention for the Protection of the Marine Environment of the N-E Atlantic
OTCs	Organotin compounds
PAHs	Polycyclic aromatic hydrocarbons
Pb	Lead
PCA	Principal component analysis
PCBs	Polychlorinated biphenyls
PhTCs	Phenyltin compounds
PMF	Penis morphogenic factor
PPAR	Peroxisome proliferator-activated receptor
PTMs	Potentially hazardous metals
PTWI	Provisional tolerable weekly intake
PVA	Population vulnerability analysis
REACH	Registration, evaluation, authorization, and restriction of chemicals
RIDDOR	Reporting of Injuries, Diseases, and Dangerous Occurrences Regulations
RXR	Retinoid X receptor
SCOEL	Scientific Committee on Occupational Exposure Limits
SDGs	Sustainable Development Goals
SLMPE	Supported liquid membrane probe extraction
Sn	Tin
TAL	Tetra-alkyl lead
TATRA	Tetramethyl arsonium
TBT	Tributyltin
TBTO	Tributyltin oxide
Ti	Titanium
TMT	Trimethyl tin
TPhT	Triphenyl tin
U	Uranium
UGT	UDP-glucuronosyltransferase
WFD	EU's Water Framework Directive
WSOC	Water-soluble organic carbon
X	Electronegativity

Chapter 1
Chemical Structure, Types, Applications, and Occurrence of OMCs

Abstract In the realm of chemistry, OMCs are an intriguing and significant topic. OMCs are classified into three categories: sigma-bonded, π-bonded, and σ- and π-bonded metal carbonyl compounds. The characteristics of OMCs as solubility, bonding type (ionic or covalent), stability, reactivity, and existence are emphasized. Applications of OMCs are highlighted. Numerous applications of OMCs are due to the distinctive array of properties and reactivities that result from the intricate interplay between metal atoms and organic ligands. The distribution of OMCs among different elements in the marine environment is tracked in this chapter, focusing on their stability. Many OMCs were identified as environmentally persistent, easily concentrated, highly toxic (often more hazardous than their original form), and degrading at a slow rate. Academics, industrial scientists, ecologists, and environmentalists who work with organometallic compounds (OMCs) will find this chapter to be a helpful guide. It covers the subject either as a stand-alone review or in the sequence of the topic area.

Keywords Composition · Metal-carbonyl bond · OMC industry · Aquatic distribution

1.1 An Overview of OMCs' Chemistry

As science and technology advance, OMCs' synthesis, properties, environmental stability, and practical applications have grown in significance to address the challenges and possibilities of our modern world. Compounds classified as organometallic have at least one bond between an organic group's metal (M) and carbon (C). An example of an OMC is titanium (IV) butoxide [$C_6H_5Ti(OC_4H_9)_3$], where Ti is coordinated with a phenyl group carbon (Abbott 1987; Meanwell 2018). In organometallic, "metallic" does not refer to "metal" but rather to any element that forms a sigma or pi bond and has an electronegativity (x) lower than carbon (2.55). Nitrogen, oxygen, sulfur, and

T. O. Said, G. M. El Zokm, *Organometallic Compounds in the Marine Environment*, https://doi.org/10.1007/978-3-031-99189-9_1

fluorine derivatives with $x = 3.00$, 3.50, 2.60, and 4.00, respectively, are therefore called organic compounds and not considered OMCs (Egorova and Ananikov 2017). Because C does not belong to the carbon of an organic group, compounds like calcium acetylide and Hg (II) fluminate are historically not regarded as OMCs (Adhikari et al. 2017). Tetra-methyl-di-arsine or $As_2(CH_3)_4$ was identified as the first OMC derivative in the mid nineteenth century, a smoked liquid with a disagreeable odor (Gafurov et al. 2021). OMCs have one or more substantial interactions between the molecular organic carbon atom with an element, including metalloids like silicon, tin, and boron, as well as transition, alkaline, and alkaline earth metals (Lagerström et al. 2017). Human activity has rapidly increased in densely populated places during the past few decades, increasing the risk of environmental degradation, especially in coastal systems. The marine environment will always undergo some changes due to any material dumped into the ocean. The alteration could be significant, minor, permanent, temporary, broad, or incredibly localized. Any change that is observable and deemed harmful is considered pollution. Measuring contamination levels in water, sediments, and organisms has been a major undertaking. However, a study of its biological impacts is usually necessary to determine whether the detected level of contamination will result in pollution. Because there are so many different ways for pollutants to enter the marine ecosystem, their effects are diverse and irregular. Monitoring the condition and processes of the aquatic ecosystem due to contaminants is just as important as controlling ocean inputs. OMCs address the topic, summarizing the latest findings and discussing their overall significance and traits in the environment. The mechanisms and pathways by which metals move through the environment require an understanding of the concentration and chemical structure of OMCs throughout the ecosystem (Craig 2023). Compounds with a straight σ- or π-carbon metal bond are their derivatives; additionally, all elements with lower electronegativity than the C atom are included in the metallic forms (Craig and Jenkins 2004). Alkali and/or alkaline earth metals are frequently the key metals implicated in environmental issues. Metalloids, such as B, Si, and Se, are also known to produce OMCs, which are used actively in the industry (Michalke et al. 2000; Andrewes et al. 2001; Craig et al. 2003; Crabtree 2009). OMCs are a fascinating and important topic of research in the discipline of chemistry because of their numerous applications in anything from materials science to catalysis and more. The majority of previous studies have solely examined total metal determination. Compared to their inorganic counterparts, OMC forms of metals are far more hazardous (Ding et al. 2017). Water contamination by metals has been identified as one of the main environmental issues for the foreseeable future. Metals will eventually pose a greater threat to the environment than organic or even radioactive wastes due to their direct emission. (Jaishankar et al. 2014). Types of OMCs depend upon the M-C bond; Zeise's salt, dibenzene and ferrocene chromium, ionic bonds, σ-σ bonds $[(CH_3)_3SnCl]$, π-π bonds, and multicenter bonds; and MeLi MeLi and $Al_2Me_6Al_2Me_6$. Research on OMC will take a multidisciplinary approach in the future, with a focus on sustainability, innovation, and the creation of new applications that have significant effects on society, technology, and science. In general, it can be said that OMCs are advantageous for society and research (Kumar et al. 2023).

A few of several types of marine pollution can take many forms, including litter, particles, noise, excess carbon dioxide, and invasive creatures, as well as garbage from industry, agriculture, and homes, which enter the water and have negative impacts. Changes in habitats, food chains, and ocean biota all have an impact on productivity patterns. Taking all of these considerations into account, it appears reasonable to argue that microbial digestion is the reliable and efficient technique to remove medicines, pesticides, pollutants, carcinogens, volatile petrochemicals, food additives, and contaminated workplace compounds from water that are alien to living systems (Datta 2023). OMCs in the environment address the topic, summarizing the latest findings and discussing their overall significance and traits. The mechanisms and pathways by which metals move through the environment require an understanding of the concentration and chemical structure of OMCs throughout the ecosystem (Craig 2023). Compounds with a σ- or π-carbon metal connection are known as OMC. Moreover, every element with less electronegativity than the carbon atom can only be included in the concept of the metallic atom (Craig and Jenkins 2004). Because of their many uses in everything from materials science to catalysis and more, OMCs are an interesting and significant area of study in the field of chemistry. Concerns about the destiny of trace elements and their effects have developed during the last two decades. OMCs are far more toxic than their inorganic equivalents (Ding et al. 2017), and a lot of environmental protection and public health organizations now want in-depth studies on OMCs. Water contamination by metals has been identified as one of the main environmental issues for the foreseeable future. Metals will eventually pose a greater threat to the environment than organic or even radioactive wastes due to their direct emission (Jaishankar et al. 2014). Research on OMC will take a multidisciplinary approach in the future, with a focus on sustainability, innovation, and the creation of new applications that have significant effects on society, technology, and science. Generally, Kumar et al. (2023) stated that OMCs benefit society and research. Carbon-metal bonds come in a variety of forms, ranging from mostly covalent to essentially ionic. The shape of the C-M bond may be somewhat influenced by the structure of any organic component in an OMC, but identifying the metal is far more crucial. The ionic character of the C-M link grows as the metal becomes more electropositive. In OMCs, an electropositive atom forms a bond with carbon. Thus, metal is positively charged, and carbon is negatively charged. Organic components are therefore always both basic and nucleophilic. OMCs can consequently function as bases as well as nucleophiles. Environmentally stable M-C bonds are found in the periodic table for germanium, tin, lead, arsenic, antimony, selenium, and mercury (Siegl and Sigel 1993); however, they may be formed mostly from industrial production wastes. It may take a long time for biological processes to degrade chemical pollutants found in water since their molecular structures are sometimes very different from those of natural substances. Therefore, even if a portion of solar energy is stored in aquatic biological systems, it won't be easily accessible for the degradation of pollutants created by humans (Gianguzza et al. 2002). The potential environmental impact of xenobiotic photoreactions in water has received far less consideration. As interest in this phenomenon has increased recently, so too have efforts to quantitatively define the kinetics of xenobiotic reactions in the environment. Certain organic and inorganic components are preferentially removed from sinking particles, causing substantial compositional changes. The rates of breakdown and transformation are determined by the chemical makeup of particular substances and their suitability as substrates for heterotrophic metabolism (Buesseler et al. 2000). Regretfully, trace metal behavior is more complicated in the real world. A variety of chemistries can take place (Morse and Luther 1999). These include coprecipitation processes, the creation of several distinct minerals, and adsorption that is essentially irreversible. The problem is made more difficult by the frequent occurrence of variations in metal valence, such as Cu (II) to Cu (I).

One of the most practical fundamental tools in contemporary inorganic chemistry is the electron rule and associated techniques for calculating the total valence electrons of transition metal complexes, especially when applied to OMC species. But this explanation is insufficient to account for the patterns and anomalies seen in real-world operations (Rasmussen 2015). When predicting the stability of metal carbonyls, this rule is effective. With certain other OMCs, though, this isn't the case. The central atom of a compound forms bonds by taking on electrons in its orbitals. Consequently, electrons can be held in the core atom's s, p, and d orbitals. The central metal atom can receive a maximum of 18 electrons from Lewis bases (Fig. 1.1). The same approach is used to get the closest noble gas configuration. Consequently, the d-electrons of transition elements must add up to 18 in addition to the number of electrons supplied by ligands.

1.2 Types of OMCs

In the lab and business, OMCs are frequently used as catalysts and intermediates. The elements are covalent because they contain a metal-carbon bond. These substances are special because, at room temperature, some of them are liquids while others are solids. This is an example of an OMC where the carbons of a benzene ring link with chromium. One example of a material in this family is ferrocene, a very stable compound with Fe sandwiched between two rings of hydrocarbons. OMCs are organic materials that contain M-C linkages. Among the metals are alkali

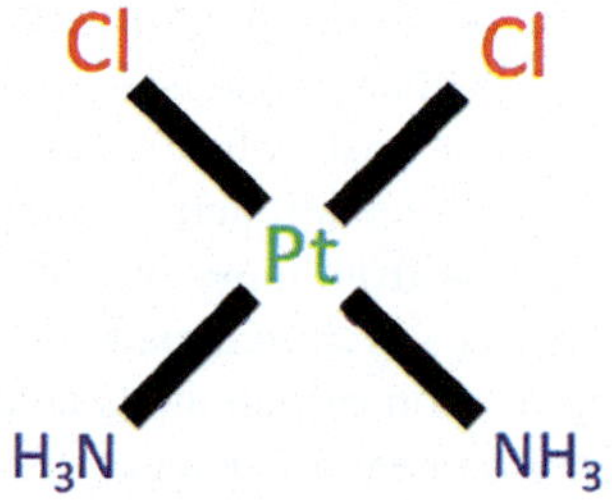

	Number of e's
Pt	10
charge	0
2 Cl	$2 \cdot 1 = 2$
2 N	$2 \cdot 2 = 4$
Total	**= 16**

Fig. 1.1 18 electron rule (https://collegedunia.com/exams/organometallic-compounds-chemistry-articleid-821)

and alkaline earth metals. Furthermore, it is occasionally known that metalloids like boron, silicon, and selenium can form OMCs used in industrial chemical reactions. Organolithium, organomagnesium, Grignard reagent (GRs), organocuprate, organozinc, organocadmium, and so forth are typical examples of OMCs (Perez 2024). The most common OMCs in aquatic environments are organotin (IV) and organomercury (II). The use of OHgC as antibacterial agents in medicine is just one of the many industrial uses for OTC (Craig and Miller 1997). Each of the OMC moieties that make up compounds containing organotin (IV) includes a covalent core element as well as a variety of chemical groups. Their general formula is $RnSnX_{(4\text{-}n)}$, where n can be any number between 1 and 4, and X can be any of the following: halide, nitrate, acetate, hydroxide, etc. (De Stefano et al. 2000):

$$qR_nSn(OH_2)_x^{z+} + pH_2O \rightarrow pH_3O^+ + (R_nSn)_q(OH)_p(OH_2)_y^{(qz-p)+}$$

1.2.1 *Classifications of OMCs*

OMCs are divided into two classes based on the type of M-C bond: (1) ionic-bonded OMCs, alkali, alkaline earth metals, lanthanides, and actinides. The primary components of ionic compounds are OMCs. These solids have no color and look like salt. When fused, they function as electrolytes. Nonpolar solvents cannot dissolve the hydrocarbons released by these hydrolysis reactions (Bhatt 2016). (2) σ-bonded OMCs: a covalent connection connects the M-C of an organic group. Low electropositive metals are the ones that create these bonds. These compounds are composed of weakly electropositive atoms, both metal and nonmetal. In these compounds, a metal atom sharing two electrons with a C forms a σ-bond, for example, π-bonded OMCs of $(C_2H_5)_4Pb$, $(Me)_2SiCl_2$, $(Me)_3SnCl$, and $(Me)_2Zn$. These complexes comprise a ligand-metal bond involving the ligand's π-electron. OMCs are often categorized as the main group, which includes metals with s- or p-blocks, such as MeLi and Al_2Me_6. The GRs, RMgX, is the most often used example. D-block metals are transition metals. $Pd(Ph_3)_4$ and R_2CuLi (Gillman's reagent) are two examples. These OMCs contain the f-block metals, actinides, and lanthanides. Uranocene is an example of an actinide and lanthanide OMC. Depending on the kind of link, OMCs can be categorized into three groups:

1. σ-bonded, in which a sigma bond connects the ligand's carbon atom and metal atom, GRs, for instance, RMgX.
2. π-bonded: these are metal compounds that contain ring compounds such as benzene, alkenes, and alkynes. The ligand's π-electrons are involved in the link that forms between the metal and ligand in these complexes (Fig. 1.2).
3. Metal bound with σ and π: metal atoms in these compounds typically have an oxidation state of zero. There are three types of carbonyls: mononuclear, bridged, and polynuclear, for example, tetracarbonyl nickel, pentacarbonyl chromium, and hexacarbonyl iron.

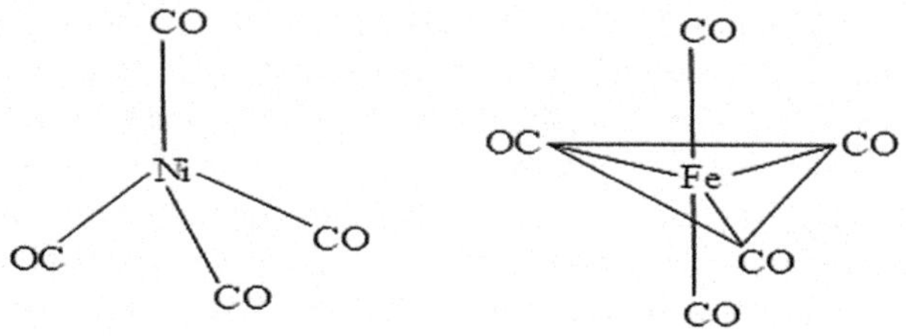

Fig. 1.2 π-π-bonded organometallic compounds (https://collegedunia.com/exams/organometallic-compounds-chemistry-articleid-821)

Craig (2003) declared that the elements of interest are usually the main group metals that are environmentally known (Tables 1.1 and 1.2). The role of OMCs in science is significant. They are widely used in chemical industries and laboratories and have many applications. The following are some instances of OMCs that the students can look up: lead, tin, organoboron, organozinc, organocadmium, organolithium, and organomagnesium.

GRs are created by applying magnesium metal to an organic halide. The organomagnesium complex needs to be stabilized by either cyclic or acyclic ethers. Water and air, which rapidly oxidize or protonolyze the reagent, are avoided when using air procedures. Ultrasound can help to create GRs; however, the reagents must remain dry (Abbott et al. 2011).

1.2.2 *Properties*

A brief explanation of a few characteristics of OMCs is the following:

1. Most OMCs are found in solid forms, especially those containing ring-structured or aromatic hydrocarbon groups.
2. The M-C bond is highly covalent.
3. They can act as reducing agents, especially when made of highly electropositive metals.
4. Highly electropositive metal compounds, like those containing lithium or sodium, are extremely volatile and burn on their own.
5. It has been discovered that many OMCs are toxic to humans (House and House 2016).

The distinguishing characteristic of OMCs is the M-C bond that is highly covalent. Many of these compounds are solid at normal temperatures, while others, such as methylcyclopentadienyl manganese tricarbonyl, are liquid. When metals with high electropositivity (like Li and Na) create OMCs, the carbon bonded to the central metal atom exhibits a carbanionic character due to the covalent nature of the link. OMCs aid in the catalysis of processes involving medicines or polymers as the target molecules. The rate of the reactions is accelerated. The toxicological behavior of OMCs differs from that of the corresponding elements' nonorganic derivatives. The impact of OMC species is typically orders of magnitude more dangerous.

Table 1.1 Stability of OMCs to oxygen[a] (Craig 2003)

Stable	Unstable[a]
Me_2Hg	$MePbX_3$
Me_4Si, $[Me_2SiO]_n$, $(Me)_nSi^{(4-n)+}$	$MeTl^+$
Me_4Ge, $Me_4Ge^{(4n)+}$, Me_6Ge_2	Me_2Zn, $MeZn^+$
Me_4Sn	Me_2Cd, $MeCd^+$
Me_4Pb	Me_3B
MeHgX (Ph and Et also stable)	Me_3Al
Me_4nSnX_n	Me_3Ga
Me_3PbX	Me_3In
Me_2PbX_2	Me_3Tl
$Me(C_5H_4)Mn(CO)_3$[b]	Me_3As[c]
$MeM_n(CO)_4L$[d]	Me_3Sb[c]
MeAsO(OH)	Me_3Bi[c]
$MeAsO(OH)_2$	Me_2AsH
Me_2S	$MeAsX_2$
Me_2Se	$MeSbX_2$
MeHgSeMe	Me_4nSnH_n[c]
$MeCoB_{12}$ (solid state)Me3SbO	Me_6Sn_2 (at RTgives $[Me_3Sn]_2O$)
$Me_3Sb(OH)$	Me_6Pb_2 (to methyl lead products)
$Me_3Sb(OH)_2$	Me_3Sb
Me_2Tl^+, Me_2Ga^+	Me_3AsO
Me_3S^+	Me_3P
Me_3Se^+	Me_4SiH_{4-n}
Me_3PO	Me_4GeH_{4-n}

[a] Variously unstable because of empty low-lying orbitals on the metal, polar M-C bonds, and lone electron pairs on the metal
[b] Gasoline additive
[c] But is stable in dilute form and detected in the environment
[d] To exemplify ligand-complexed transition metal organometallics. Many of these synthetic compounds are oxygen-stable, but none have been found in the natural environment
[e] At room temperature in bulk as against rapid (seconds, minutes) oxidation. Assume similar but lesser environmental stability for ethyls

Therefore, it is essential to differentiate between the many ways that elements and compounds bind. They can be found in the natural world or have been brought in by human pollution. OMC species are created spontaneously by chemical or biological methylation of inorganic substrates (Craig and Glocking 1988). However, they may be formed mostly from industrial production wastes. Relevant to that, OMCs are introduced into the environment, as biocides in agriculture, or using petrol additives, polymer stabilizers, or antifouling paints (Craig 1986). Figure 1.3 gives an overview of OMC examples (Leadbeater 2007).

Sn, Hg, and Pb are the basic components in the formation of OMCs. The use of copper phthalocyanine antifouling paint has been used extensively over the last 50 years. The mollusks develop high immunity to that antifouling compound. That made the use of mercury butadiene on a large scale widespread after the failure of

Table 1.2 Stability of OMC in water (Craig 2003)

OMC	Stability, comments
R_2Hg, R_4Sn, R_4Pb	Only slightly soluble, stable, and diffuse to the atmosphere. Higher alkyls are less stable and less volatile. Species are hydrophobic and variously volatile
MeHgX	Only slightly soluble, depending on X
$Me_nSn^{(4-n)+}$	Soluble, methyltin units are stable but made hexa- and pentacoordinate by H_2O, OH^-. Species are solvated and partly hydrolyzed to various hydroxo species. At high pH, polynuclear bridged hydroxo species form for $(CH_3)_2Sn^{++}$
Me_3Pb^+	Soluble, hydrolysis as methyl tins above. Also dismutases to $(CH_3)_4Pb$ and $(CH_3)_2Pb^{++}$ at 20 °C
Me_2Pb^{++}	Soluble as for $(CH_3)_3Pb^+$ above. Disproportionates to $(CH_3)_3Pb^+$ and CH_3. These reactions cause eventual total loss of $(CH_3)_3Pb^+$ and $(CH_3)_2Pb^{++}$ from water
Me_2As^+	Hydrolyzes to $(CH_3)_2$ AsOH and then to slightly soluble $[(CH_3)_2As]_2O$
$MeAs^{++}$	Hydrolyzes to $CH_3As(OH)_2$ and then to soluble $(CH_3AsO)_n$
$Me_2AsO(OH)$	Stable and soluble (330 g dm^{-3}). Acidic pKa = 6.27, i.e., cacodylic acid, dimethlyarsonic acid. Detected in oceans
$MeAsO(OH)_2$	Stable and soluble. Strong acid pK_1 = 3:6, pK_2 = 8:3—methylarsinic acid. Detected in oceans
Me_3As^+, Me_3Se^+	Stable and slightly soluble
$MenSICl_4$	Hydrolyze and condense, but methyl silicon groupings are retained
$Me_2Ge^{(4-n)+}$	Stable and soluble; they have been discovered in oceans. Hydrolyze but $(CH_3)_n$ Ge moiety preserved
Me_2Tl^+	Very stable and soluble but has not been detected as a natural environment product
Me_3AsO, Me_3SbO	Stable and soluble
MeAsH, CH_3AsH_2	Insoluble, diffuse to the atmosphere, air is unstable
MeSbO(OH)	Stable and soluble. Detected in oceans
$CH_3SbO(OH)_2$	Stable and soluble. Detected in oceans

Solubility refers to air-free distilled water, with no complexing ligands. The range of solubility is from mg dm^{-3} to g dm^{-3}. Other species stable and insoluble, R_4Si, $(R_2SiO)_n$ $H_3HgSeCH$, most PhHg derivatives, Me_2S, Me_2Se, Me_4Ge, Me_3B and unstable, $MePb^+$ (has not been deteeted in the environment), R_2Zn, R_2Cd, R_3AI, R_3Ga, Me_6Sn_2, Me_6Pb, $MeTl^{+2}$, $MeCd^+$, Me_2Cd, Me_2Sb+, $MeSb^{+2}$

copper phthalocyanine in protecting outdoor surfaces. However, the use of mercury causes many toxicity problems for biota other than mollusks. OHgCs are also used in agriculture and industrial activities (Lobinski et al. 1994). However, the primary purpose of OPbCs is to prevent knocking in gasoline engines (Jantzen and Prange 1995).

The use of OTCs has increased recently. Tin species that are mono- and dialkylated are employed as catalysts or to increase the heat and light stability of polymers. Tri-OTCs are employed in the maritime and agricultural sectors. The uses of OMCs as intermediates that are produced from transition OMCs are used as homogeneous catalysts for many processes. They are used in the field of

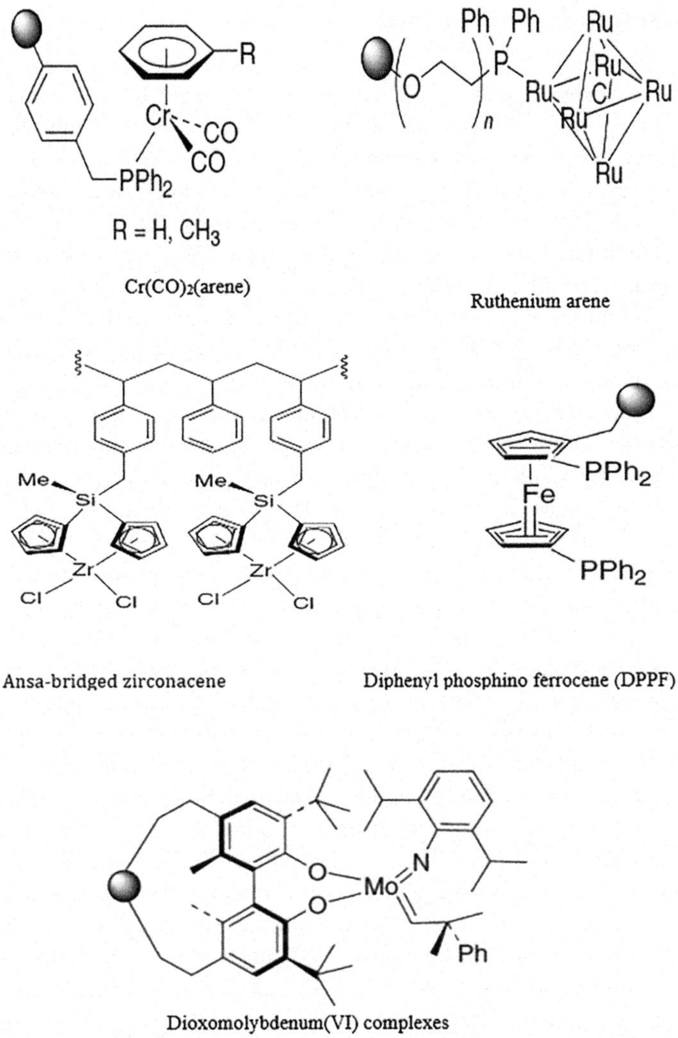

Fig. 1.3 OMCs examples (Leadbeater 2007)

medicine. For example, in contemporary surgery, Si rubbers are utilized as replacement body components. Metal purification is another application for OMCs. Numerous organic chemicals, including OMgC and OLiC, are synthesized using them. In agriculture, they are also used to treat seeds, preventing infections in young plants. Among the essential characteristics displayed by OMCs are they exist mostly in solid form but also liquid form, particularly when accompanied by ring-structured hydrocarbon groups. Electropositive metals form OMCs, which function as reducing agents. Combustion: The compounds made from highly electropositive metals, including Na and Li, are very flammable and have the potential to burn on their own.

1.2.3 Uses and Applications

OMCs are very significant; new catalysts and reagents for synthesis have been developed as a result of advancements in this field. (1) Wilkinson's $[RhCl(PPh_3)_3]$ uniform catalysis is one example of an OMC that catalyzes the hydrogenation reactions of alkenes. (2) A variety of organic compounds are synthesized using OLiC and OMgC. (3) Heterogeneous catalysis: Low-temperature alkane polymerization can also be facilitated by combining an OMC molecule. (4) Agriculture: To avoid infecting immature plants, OMCs, like C_2H_5HgCl, are sprayed on seeds. (5) Medicine: The mainstay of treatment for syphilis is a range of organoarsenic compounds. In contemporary surgery, silicone rubbers are utilized as replacement body parts. In summary, this article leads us to the conclusion that OMCs are extensively employed in the synthesis of numerous organic compounds, medications, and agricultural goods, among other things. However, some environmental problems are also caused by these chemicals, as some of them are toxic. In electrochemistry, OMCs are used in many different contexts. Their low oxidation state of metal atoms and structural variation make them attractive candidates for future treatments (Parham et al. 2023). Electrochemical chemo−/biosensors employ metal-containing organometallic polymers for sensing purposes through a variety of transduction methods (Feng et al. 2015). To improve conductivity without changing the electrical structure, graphene nanoplatelets' surfaces can also be connected via organometallic chemistry (Chen et al. 2016). Furthermore, thermodynamic, kinetic, and mechanistic information has been obtained in electrochemical systems by the use of organometallic compounds (Blakemore 2022). These substances may find use in wastewater treatment, renewable energy, and medication delivery (Parham et al. 2023). All things considered, OMCs are important for several different electrochemical applications. Rigid polymers known as silicone resins are created by hydrolysis of R_2SiCl_2 combined with a significant amount of $RSiCl_3$. Using organosilicon esters $RSi(OR')_3$ to treat hydroxyl-terminated polysiloxanes $HO(SiMe_2O)$ nH gives them greater structural control. They are widely cross-linked and used to create glass-cloth laminates, which are electrically insulating materials. When trifunctional organ silicon derivatives $RSiX_3$ are hydrolyzed alone, amorphous, strongly cross-linked polymers are typically the result. Crystalline oligomers have been isolated in a few instances. Si forms the basis of the octamer $(MeSiO_{1.5})_8$, some of which are cage molecules. OSiC is not harmful at all. Human surgical implants have been created using silicone rubber. In addition to their usage as medications, OAsC have been employed as fungicides, bactericides, and herbicides. Less hazardous tin compounds are replacing them. Every arsenic compound ought to be handled as toxic. The halides result in severe burns to the skin that are hard to recover from. When several of these compounds were created as chemical warfare agents during World War I, this trait was employed maliciously. A mixture of chlorovinyl and ethyne derivatives is produced from $AsCl_3$ with $AlCl_3$ present (Atwood 1985). Trimethyl bismuth is spontaneously inflammable in the air. There are not many different types of organobismuth compounds. The GRs method can be used to prepare the trialkyls and triaryl. Trimethylbismuth spontaneously ignites in the atmosphere. It was previously believed that organotin biocides were not harmful to mammals; however, new research has cast doubt on this. Concentrated solutions irritate the skin and eyes. In the environment, TBT easily breaks down into inorganic tin (II) compounds (Powell 1988).

1.3 OMC Chemotherapeutics with Exceptional Stability

Significantly, catalysis and biosensing applications have recently extended to pharmaceutical chemistry and all common types of OMCs. OMCs offer a great deal of potential for creating new classes of pharmacological compounds with possibly unique metal-specific mechanisms of action (Fig. 1.4).

The development of metallodrugs for cancer treatment began with the introduction of the clinically approved medication cisplatin. However, because cancer cells are resistant to several drugs and some side effects persist, there hasn't been much progress in this area up to this point (Zaki et al. 2019). To prevent cancer cells from becoming resistant to the treatments, OMCs have a different molecular target than traditional metal-based drugs, which only target DNA. While some OMC medicines target mitochondria and the endoplasmic reticulum, others work by inhibiting enzymes like thioredoxin reductase (TrRx) or targeting specific proteins. In that study, they primarily address the mechanisms of action of Ru, Ti, Au, Fe, and Os. OMCs, as well as how hyphenated techniques, enhance their therapeutic potential against diverse cancer subtypes. They also discuss how structure-reactivity relationships could assist drugs combat cancer in vivo and in vitro for human health. Many researchers are searching for novel cancer therapy agents with enhanced activity and reduced toxicity because of the current focus on the development of chemotherapeutics (Allardyce et al. 2005; Jaouen 2006; Jaouen and Dyson 2007; Hartinger et al. 2012). For centuries, OMCs have been studied as catalysts since Kopf and Kopf-Maier's groundbreaking study on titanocene. Nonetheless, they are presently being researched to find new, perhaps anticancer drugs (Köpf-Maier 1990, 1999; Köpf-Maier and Kopf 1994). Following the entry of multiple titanocene-based organometallic complexes into clinical trials, it was discovered that their mechanisms of action differed from those of the medication cisplatin, which was approved

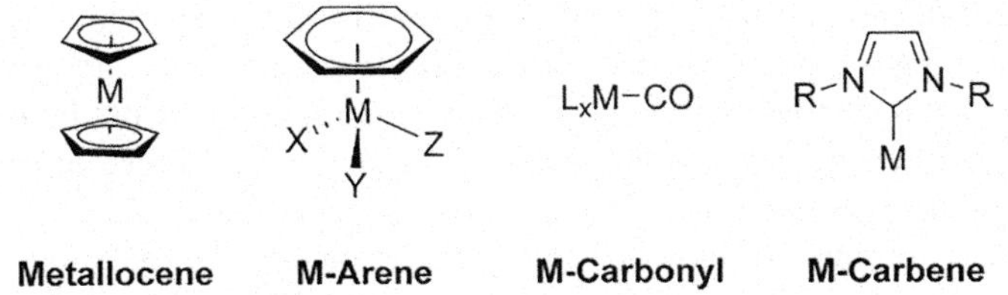

Fig. 1.4 Organometallic classes used in medicinal chemistry (Gasser et al. 2011)

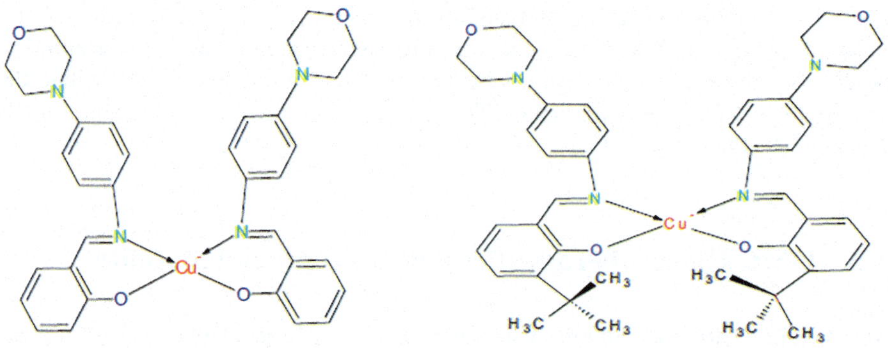

Fig. 1.5 Cu(II)-derived complexes of morpholine-derived Schiff base ligands were studied for their antitumor activity against human breast cancer cells (Shekhar et al. 2022)

for use in medicine (Christodoulou et al. 1998; Cini et al. 2013). Researchers have since concentrated more on developing OMC anticancer medication. As a result, using OMCs opens up a lot of possibilities for sensible medication design that has various mechanisms of action within tumor cells, improved cellular absorption, and fewer undesirable side effects. These drugs work differently from cisplatin, are less toxic, and bind with the DNA of the biological target by ligand hydrolysis (Melchart and Sadler 2005; Yan et al. 2005; Schfer et al. 2007; Strohfeldt and Tacke 2008; Ndagi et al. 2017). Cisplatin and its analogs brought significant attention to the topic of anticancer metal-based drugs. Because traditional anticancer drugs have inherent resistance and toxicity, researchers are focusing on developing chemotherapeutic drugs by using unique metal ions and novel approaches. The design of chemotherapy agents is challenging, and to create the best antitumor medications, we must consider both the action on tumor cells and cytotoxic activity. Furthermore, suitable strategies based on metals, target selection, and drug interaction with biomolecules are required for the administration of pharmaceuticals. It is essential to comprehend how metal ion behavior and ligand reactivity relate to target site identification or the structure-reactivity relationship, applying sophisticated methods and computational molecular orbital computations.

OMC medicines have more than just DNA as their primary target; they also work through additional targets and mechanisms. Certain agents induce stress in the endoplasmic reticulum and mitochondria, while other agents specifically target proteins and different enzymes, including thioredoxin reductases (TrRxs), which ultimately result in cellular death. Scientists in this discipline are primarily concerned with synthesizing potent anticancer medicines that specifically target tumor cells. This can be achieved by employing novel metal ions and altering the ligand structure. These will create new opportunities for the future development of OMC chemotherapeutics. Shekhar et al. (2022) discussed metallodrug (OMC Schiff base) potential and ultimate destiny as antibacterial or anticancer agents (Figs. 1.5 and 1.6).

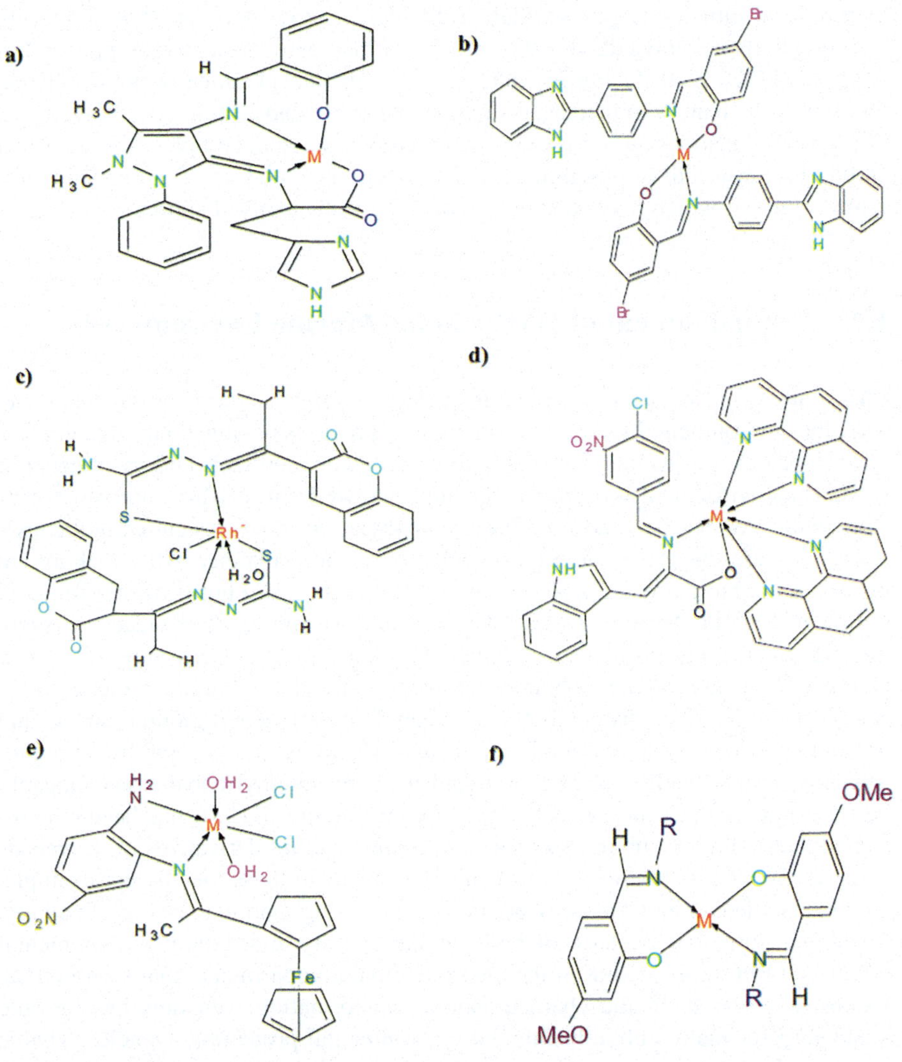

Fig. 1.6 A series of metal complexes (M = Co(II), Ni(II), Rh(III), Cu(II), Zn(II), and VO(IV)) as antibacterial (Shekhar et al. 2022)

Finding alternatives platinum-based chemotherapy for cancer patients' regimens is critically important (Gasser et al. 2011). Despite their outstanding accomplishments, there are two major drawbacks to these platinum complexes: they have detrimental side effects, such as nephrotoxicity, and are ineffective against platinum-resistant cancers. The latter disadvantage results from the fact that these medications' ultimate target is widely available: since DNA is found in every cell, it is commonly acknowledged that Pt anticancer medications target it (Lippert 1999; Wang and Lippard 2005). Furthermore, due to its distinct chemical structure, cisplatin offers few chances to increase its tumor selectivity and lessen unwanted effects. This essential distinction from "classical coordination metal complexes" means that OMCs have a wealth of possibilities for designing new classes of therapeutic molecules, some of which may have metal-specific mechanisms of action.

1.4 Apportionment of OMCs in the Marine Environment

OMCs are found naturally as a result of their creation or introduction. They can enter the environment directly by being used as products having environmental features (biocides, for instance) or indirectly as polymer stabilizers and gasoline additives. The environment is significantly impacted by OMCs, which are substances that contain As, Hg, Sn, and Pb. Through methylation processes called biomethylation, they are created in an environmentally friendly manner. After a thorough investigation, biomethylation activity was discovered in soil but is mostly found in sediments found in harbors, rivers, lakes, and oceans, among other places. As such, the biological forms of tin, lead, and mercury are of concern. These substances may be found in the environment naturally or due to industrial pollution. A metal (oid)'s chemical and physical characteristics are altered when a methyl group is added, and this in turn affects toxicity. Metals can be methylated by microscopic living organisms, lower algae, and fungi. Higher organisms with a methyl-cobalt bond appear to be incapable of methylating real metals, except vitamin B12. It has been demonstrated that numerous higher species can produce methyl derivatives of the metalloids As, Se, and Te; therefore, this is not the case with these metals. For example, methyl arsenic can be formed by organisms, including mammals and marine biota (including humans). Because of both human activity and natural environmental products, OMCs are present in the environmental compartments that surround us. Extensive research on the characteristics, transformation pathways has become mandatory for many public health and environmental protection agencies. This is because OMC trace metal species are often more problematic than their parent compounds (Lee and Oshima 1998; Ding et al. 2017). These compounds meet the criteria for possessing a metalloid-carbon bond or M-C. Covalent bonds between soft ligands and soft acid metals are commonly seen in them. Their cycles and routes between different environmental phases and compartments are mediated by physical, chemical, and biological processes (USEPA 2007). Their interactions with other chemicals and ambient compartment biota dictate their destiny and pathways in the environment. The fact that OMCs can impede their functionality once they accumulate in various environmental compartments adds to the concerns surrounding them that impact the environment, vegetation, and other micro- and microorganisms, as well as contaminate the human food chain (Hu et al. 2016; Wu et al. 2016). Adsorbed inorganic species can undergo a range of biogeochemical reactions, including oxidation, reduction, methylation, alkylation, and biomethylation, after being hosted by metal oxides and organic compounds in the soil. As a result, highly mobile organic species are produced, which can then combine in living things' body tissues; hence, those organisms will absorb and accumulate them (Challa et al. 2008; Selin 2009). Furthermore, the formation of bonds between metals or metalloids and methyl, ethyl, or alkyl groups alters the solubility or volatility of the latter, which may significantly affect their fate, routes, and life cycle within and between related environmental compartments. Since the pollutant's rate of leaching is largely dependent on its unique geochemical properties, local conditions found in the ecosystem, as well as microbial activity (Craig et al. 2003; Clarckson and Magos 2006), OMCs are normally occurring elements of the Earth's crust that are produced by biogeochemical processes. Mining, agriculture, urbanization, and industry are just a few of the numerous anthropogenic sources that have significantly increased their inputs in different environmental compartments during the past 50 years, making their presence a major worldwide concern. Normally, they are found in small amounts in natural environmental components like soil or water (Aran et al. 2001; Zhaou et al. 2011; Lu et al. 2012; Pan et al. 2016). OMC distribution throughout all compartments is greatly influenced by soil, which frequently serves as a basic tank for OMCs resulting from their high absorption capacity (Biester et al. 2002; Reis et al. 2010). Metals can be found in both terrestrial and aquatic habitats. These forms include elemental forms, OMCs, and salts. Since most metals are in the form of insoluble minerals, these species will be difficult for plants to transfer or absorb in the near future (Henriques et al. 2013).

OMCs, especially volatile compounds, have been shown to be harmful to humans, taking into consideration substances that have an impact on the habitat and have the activity to endanger people health and biota. As a result, we only included organic forms of Pb, Hg, and Sn in this chapter. These compounds can be present naturally or as a result of pollution from industry. The majority of the studies conducted on OMCs were conducted in aquatic environments. Numerous investigations of OMC contamination of soil or air were conducted at industrial sites. Similar to this, studies on the bioconcentration factor (BCF) of OMCs were conducted on a variety of aquatic organisms; however, little information on the potential uptake of them by vegetables has been published in the literature. A sophisticated sample preparation method and extremely sensitive detection devices are necessary for the study of OMCs (Harrison and Rapsomanikis 1989). Hg is very hazardous to health and the worldwide habitats (Selin 2009; Henriques et al. 2013). Over the past few centuries, anthropogenic inputs have disrupted its biogeochemical cycle. The coal power plant and chlor-alkali industries, as well as other human undertakings like industry, chlorinated pesticides, and dyes, have all made use of the OHgCs (Biester et al. 2002). Severe environmental contamination was ultimately caused by emissions and inputs from the aforementioned industrial operations, burning fossil fuels, and trash disposal (Briant et al. 2017). Although Hg and its related compounds are recognized to be harmful to human health, the level of toxicity varies substantially depending on the metal's chemical state. For example, inorganic Hg salts are believed to be less dangerous than organic ones like methylmercury or dialkyl mercury (Caballero et al. 2017). Metal speciation reveals the existence of several types of metals, their toxicity, and destiny in different habitats. These metal forms include those that occur in their solids or have coprecipitated in major metal solids (USEPA 2007). It becomes vital to specify Hg regarding the toxicological effects of its species (Morita et al. 1998). Lindqvist et al. (1984) identified three categories for Hg species: (1) volatile element, (2) reactive (aerosol particles, complexes, and organic acids), and (3) nonreactive and other OHgC (Ser et al. 2017). The properties of Hg and its related compounds are also influenced by their speciation, in addition to their toxicity (USEPA 2007). The following organic forms of Hg have been identified as significant from a toxicological and hazard perspective and as being present in the environment: dimethylmercury $(CH_3)_2Hg$ (Siegler et al. 1999), phenylmercury ($C_6H_5Hg^+$), methylmercury (CH_3Hg^+), and ethylmercury ($C_2H_5Hg^+$) (Zimmermann et al. 2013; Marques et al. 2016). There are various ways that organic lead enters the environment, but the majority of the time, it is through the exhaust of cars and the petroleum sector into the air, where it contaminates soil and water sources (Zuniga et al. 2009). Tetramethyl and tetraethyl lead, two organic forms of lead, are commonly employed as fuel antiknocking additives. Lead gasoline is still used in several nations worldwide, despite a significant decline in use over the last 10 years (ACFA 2010).

The highest amount of alkyl Pb is still found in aviation fuel. In the past, the primary sources of alkyl Pb in the environment were spillage from indoor and outdoor fuel facilities. Because of its lipophilicity, tetra-alkyl Pb is one of the alkyl Pb compounds that biological organisms may readily absorb. The type of chemical, length of exposure, and type of organism all affect how well they are absorbed (Gallert and Winter 2002; Zuniga et al. 2009). Research on human toxicity has demonstrated that long-term cumulative exposure to organic Pb is more detrimental than inorganic Pb exposure (Blakey et al. 1992). The sequence in which the toxicity of alkyl Pb compounds decreases with a decreasing number of ethyl or methyl moieties or with a decreasing number of carbon atoms (ethyl lead → methyl lead) is $R_4Pb > R_3Pb^+ > R_2Pb^{2+}$, where R is either $-CH_3$ or $-C_2H_5$ (Pyrzynska 1996; Gallert and Winter 2002). When carbons are directly bound to tin, the resulting OTCs are known as OMCs (Gao et al. 2017). Plastic, catalysts, preservatives, paints, and biocide products (fungicides, insecticides, and bactericides) in agriculture are only a few of the industrial uses for OMC forms of Sn as active agents (Cole et al. 2015). The use of OTCs as antifoulants is prohibited because of the serious harm they cause to aquatic life (Birchenough et al. 2002). Furthermore, the usage of TBT and TPhT in industry has caused a lot of concern since they can have significant negative impacts on organisms, as dissolved in the surroundings, even at nanogram levels (Nakanishi 2008). These OMCs are categorized as obesogens, hepatotoxins, and teratogenic substances, as well as impair the capacity of mammals' reproduction (Zuo et al. 2011; Cole et al. 2015). Many sites are still used today for grazing or agriculture due to their extensive historical use, inappropriate disposal, and lack of assessment of potentially contaminated sites from previous activities. Exposure to contaminants such as Pb, Hg, or Sn could occur today if the distribution and speciation of these substances in soils were not properly evaluated, as well as the risks they pose to humans and animals. Recent developments in the techniques for recognizing and categorizing environmental OMCs are summarized. Both the usage of plasma spectrometry as a detector and HPLC as a separation method are growing in popularity (Chau and Wong 1991). There have been reports of new methods to get over the previously challenging HPLC to furnace AAS interface. Additionally, new methods for derivatizing ionic OMC species for gas chromatography have been developed. The sensitivity and specificity of tandem analytical methods are increasing in sophistication.

1.4.1 Seawater

Numerous metals and metallic compounds, some of which are hazardous to life, are found in the marine environment. Each metal's chemistry dictates how it reacts to the ligands in the aquatic medium. The following equation can broadly describe this reaction:

$$A + :B \rightarrow A:B$$

To differentiate between the stability of species A and B, Pearson (1968a, b) divided donors and acceptors into "soft" and "hard" categories based on experimental data (Table 1.3). Mobility, polarizability, and electronegativity are the main factors used to categorize trace metal species in aquatic systems as hard or soft. Low electronegativity and low polarizability are characteristics of a hard acceptor. For a soft donor atom, the conservation holds for a large positive charge density (such as a small radius and a high oxidation state). Because electron pair acceptors are frequently classified as acids and electron pair donors as bases, HSAB describes the favored bonding between hard species as opposed to soft ones. The fact that some elements are observed naturally as sulfides, others as oxides or carbonate ores may be one of the earliest instances of the HSAB principle in action. On the other hand, soft bases like S^{2-} are preferred by softer acids like Hg_2^{2+}, Hg^{2+}, or Pb^{2+}. There are three distinct chemical states of heavy metals in seawater: particulate, colloidal, and dissolved. According to Sigel and Sigel (1984), metal speciation can reveal details

Table 1.3 Acceptor and donor classification (Pearson 1968a, b)

Acceptors		
Hard	Intermediate	Soft
H^+, Na^+, K^+, Be^{2+}, Mg^{2+} Ca^{2+}, Mn^{2+}, Al^{3+}, Cr^{3+} Co^{3+}, Fe^{3+}, As^{3+}	Fe^{2+}, Co^{2+}, Ni^{2+}, Cu^{2+}, Zn^{2+}, Pb^{2+}	Cu^+, Ag^+, Au^+, Ti^+, Hg_2^{2+}, Pd^{2+}, Cd^{2+}, Pt^{2+}, Hg^{2+}, CH_3Hg^+
Donors		
Hard	Intermediate	Soft
H_2O, OH^-, F^-, Cl^-, PO_4^{3-}, SO_4^{2-}, CO_3^{2-}, O_2^-	Br^-, NO_2^-, SO_3^{2-}	SH^-, S_2^-, RS^-, CN^-, SCN^-, CO_2, R_2S, RSH, RS^-

Note: R alkyl or aryl

about the intricate relationships that exist between biological components, sediment, particulate matter, and dissolved substances in aquatic environments. The NOAA reports from May 2008 indicate that there is a trend toward less pollution in coastal waters. The report, which is based on research conducted over two decades (1986–2005), demonstrates that environmental regulations passed in the 1970s are helping to lower the overall levels of pollutants in American coastal waters. However, the study did draw attention to persistent concerns regarding elevated levels of metals and organic contaminants near industrial zones and coastal cities (Kimbrough et al. 2008).

The sources of each metal, the chemistry of its ions in the most common oxidation state, the way the ions interact with the different water components, and the concentration of each metal are mostly determined by the many parameters (pH, dissolved oxygen, oxidation reduction potential, salinity, and temperature). Zinc hydrolyzes between pH 7.00 and pH 7.50 to produce the comparatively stable $Zn(OH)_2$, which is generated at pH > 8.00. Zinc is found as divalent zinc at a hydrogen ion concentration of 6.70, where it can combine with organic materials and sorb onto suspended mineral colloids. Zinc chloride complexes don't form until concentrations are higher than 890×10^2 mg L^{-1}. While organo Zn compounds are more stable due to the donating atoms of N and S in the ligand, they also have a strong propensity to adsorb to clay minerals and suspended organic matter or precipitate with iron or manganese oxides, which enriches sediments and extracts zinc from the water column. Many different kinds of inorganic and organic ligands form complexes with water-soluble zinc. In acidic freshwater conditions, up to 50% of the total zinc can be found in inorganic forms such as zinc carbonate, zinc hydroxyl carbonate, or zinc silicate. The majority of bound zinc in alkaline fresh waters is adsorbed onto organic and inorganic colloidal particles. Most fresh waters have pH values within the range needed for heavy metals to bind to particles. The majority of zinc can be found either adsorbed or desorbed, depending on a pH shift of 0.500. Zinc was found in both stable inorganic form and ionic species at pH 6.00–6.10 (Gaiero et al. 2002). There have been reports of increases in the dissolved and suspended zinc fractions in estuary water in the area where fresh and brackish waters interact. The increases were attributed to zinc's greater residence time in the estuary compound as opposed to its residence time in fresh water. High temperatures, low hardness, and low pH all promote zinc's easier dissolution (Malle 1992). According to Buck and Bruland (2005), copper in waters is either adsorbed onto particulate matter or complexed with different ligands. Only trace amounts of copper are found in natural waters as the free aqua ion "Cu^{2+}." The most significant inorganic ligands are hydroxide, carbonate, nitrate, sulfate, and chloride in seawater. A significant amount of dissolved copper is frequently organically complexed due to the strong binding of copper to fulvic and humic acids as well as other organic compounds (Gordon et al. 2000). As a function of pH (3.50, 4.00, and 4.50) and Cu/C, the relationship between Cu and dissolved organic matter (DOM) was examined at pH (3.50, 4.00, and 4.50). At pH 4.50, the dissolved organically bound fraction peaked; however, as the Cu/C ratio increased, this fraction decreased, and a trend was noted at all pH levels (Nierop et al. 2002). Researching the solubility of dissolved organic carbon (DOC) and fluvic acid (FA)-metal complexes concerning pH, metal ion concentrations, and FA concentrations revealed that the stability of water-soluble organic carbon (WSOC) complexes increased with pH but less than that of DOC and hard acid complexes. When the concentration of total copper ions increased at pH 4.00, the stability constant shrank (Cao et al. 2004). About 90% of the $\sum$Cu is linked to humic materials in fresh water, while only 10% is bound by them in seawater because of the high concentrations of calcium and magnesium, which displace copper from humic materials. According to Cances et al. (2003), the dissolved organic matter controls copper speciation, and the mobility of that organic matter controls copper transport. In most fluids with pH values between 4.00 and 7.00, Mn(II) is predominant and mostly associated with carbonate, which has a relatively poor solubility (Schaanning et al. 1988). Manganese oxide equilibria, which involve the conversion of manganese into different oxidation states, can regulate the solubility of Mn(II). The formation of the poorly soluble sulfide tends to control the fate of manganese in extremely reduced water. It has been discovered that methylcyclopentadienyl manganese tricarbonyl (MMT), is persistent in naturally occurring aquatic and soil environments when sunlight isn't present with a propensity to sediment. MMT photodegrades quickly in light, yielding products such as manganese carbonyl, which easily oxidizes to manganese teroxide. Together with both organic and inorganic ligands, Ni(II) forms complexes. Carbonates, phosphates, sulfates, and halides are examples of inorganic complexes. Strong complexes with nickel are formed by organic ligands containing donor atoms of oxygen, nitrogen, and sulfur. The destiny of Ni in fresh and salt water is affected by many variables, including pH, ligands present, adsorption on solid matter, and others (Nriagu 1980). The most prevalent form of Ni in natural waters is the divalent ion, which is represented by the form $Ni(H_2O)^{2+}$ and has a pH range of 5.00 to 9.00 (Skeaff and King 2000). In fresh waters, Cr^{+6} is reduced Cr^{+3} and then sorption to sediment particles (Ambasht and Ambasht 2001). According to Loyaux-Lawniczak et al. (2001), Cr^{+3} is comparatively insoluble and frequently coprecipitates with Fe-oxyhydroxides, whereas Cr^{+6} is highly soluble and toxic. Therefore, Cr mobility in aqueous solutions is controlled by redox conditions and reactions. When household waste is poured into rivers, H_2S is created, and the water's dissolved O_2 content suddenly decreases. As a result, the percentage of total Cr in solution present as Cr^{+6} dropped from 87% to 34%. According to Florence (1982), there were 45.0–65.0% organic species, 25.0–40.0% Cr^{+6}, and 10.0–20.0% inorganic species in seawater. Monomethylarsonic and dimethyl arsinic acids, arsenate, and arsenite are the dissolved forms of arsenic in the water column. In certain cases, As^{+3} and As^{+5} species can switch oxidation states based on biological processes and pH. Certain species of arsenic may exhibit different behaviors in different environments due to their preference for clay minerals and organic matter. Important changes that regulate the mobilization and subsequent dispersion of arsenicals include methylation and demethylation reactions (Nriagu 1994). The way that arsenic travels through water and how it interacts with other substances is determined by its chemical composition. Quaghebeur et al. (2005) claimed that arsenic can be absorbed from water by clays, Fe oxides, $Al(OH)_3$, Mn compounds, and organic molecules.

1.4.2 Sediment

Organic chemicals found in the sediment are essential for the transformation of heavy metals and are frequently found in considerable particle form. For example, in certain rivers and lakes, the majority of the silt is usually composed of heavy metals bound to organic substances. Furthermore, the mobility of metals in sediment is typically directly impacted by the solubility of organic materials. Amina et al. (1999) stated that forming soluble metal complexes with dissolved organic compounds would increase the mobility of the metal ions, while complexing metal ions within soluble organic compounds can often significantly reduce their mobility. Mobility, activity, and bioavailability are influenced by a variety of physicochemical variables (Filgueiras et al. 2002; Morillo et al. 2002; Sekhar et al. 2003). In general, low redox potential and pH levels allow heavy metals to be easily mobilized into the solution. The exposure of sulfide minerals in sediments to atmospheric oxygen lowers pH levels and causes heavy metals that are bound to the sediments to undergo new distribution and transformation. Due to active sites competing with one another, increased salinity can also result in dissolution (Joseph 2002). Heavy metal accumulation in sediments may turn into a secondary cause of water contamination once the environment has changed (Cheung et al. 2003). Thus, one of the most important factors in assessing the risk to aquatic habitats is figuring out how contaminated sediments are with metals (atomic weight >20).

1.4.3 Aquatic Organisms

It is well known that biological systems depend on specific metal ions. They are commonly known as essential elements. Furthermore, it was discovered that minute amounts of elements are necessary for the existence of beings. However, several additional metals are hazardous (Chen et al. 2001). When an ingredient is regularly found to be present in every biological system, it is considered essential. Since deficiency symptoms are associated with the removal or depletion of an element and go away when it is reintroduced into the biological system, they ought to be linked to a particular biochemical issue. Heavy metals are necessary for creatures; however, these metals become hazardous as their concentration surpasses what is needed for the proper nutritional response by factors that can range from 40 to 200 times. As a result, when an organism cannot withstand an increased concentration of metal through direct use, storage, or excretion, metal toxicity results. In light of this, metal toxicity happens when an organism cannot withstand an increase in metal concentration through direct use, storage, or excretion (Guven et al. 1999). The physicochemical form of heavy metals dictates their toxicity to aquatic plants and animals in aquatic systems; however, this is irrelevant. For instance, while Cu ions are extremely toxic, Cu bound to naturally occurring organic matter is safe. After settling and assimilating into sediments, particulate matter releases harmful trace elements into aquatic systems. The effects of various toxicants will also rely on the overall quality and condition of the water, as well as the amount, kind, and concentration of other toxicants that are present. The concentration of the toxicants in the water, their form (ionic, complexed, or organic), and even the kind of organism they impact are some examples of these variables. Both direct absorption from the water and uptake through food are two ways that aquatic creatures might acquire critical metals and thus hazardous ones as well. The relative significance of each of these routes is significantly influenced by biological and environmental factors. The diversity in size, sex, and age of individuals has been a major cause of variation in trace metal tissue concentrations in aquatic organisms. An organism's age, both individually and with its species, can influence whether heavy metal enrichment occurs over an extended time. The species employed in all research studying the association between metal concentration and age in aquatic organisms came from the same biotope. The reason this is required is to rule out any local factors that could impact the quantity of metal in a given organism (Al-Yousuf et al. 2000; Pourang and Amini 2001; Farkas et al. 2003).

Studies on the topic have shown that age affects the overall Hg content of fish. Higher levels of MeHg can be accumulated by older organisms than by quickly growing younger ones (Palmer and Presley 1993). However, this continuous correlation between Hg and animal size might not apply to other metals because they do not show a distinct and steady pattern for a link with animal sizes (McCoy et al. 1995; Al-Yousuf et al. 2000; Nussey et al. 2000). Storelli and Marcootrigiano (2002) looked at the connection between Se and Hg in *Galeus melastomus* from the Mediterranean Sea. The researchers discovered that the regression slope of Hg in the liver on weight was positive, equal to 0.53 at $p < 0.0001$. This implies that either Hg intake rises with age or Hg elimination rates fall, maybe as a result of declining demethylating effectiveness. Methyl mercury biotransformation in animals primarily occurs in the liver, and after a component is expelled into bile, the main route for methylmercury removal is through feces (Boening 2000).

TBT and their degraded levels in bivalves *Mytilus edulis* from Nuuk town, Greenland, had approximately 1 pg kg^{-1} wet wt, estimated as Sn, which was lower than that discovered in Iceland and the Faeroe Islands. TBT concentrations in sediments ranged from <1.00 to 171 pg kg^{-1} dry wt, estimated as Sn, which is equivalent to levels found in Europe (Jacobsen and Asmund 2000; Jacobsen and Strand 2000). $\sum$Sn was also determined in pelagic fish (*D. sargaus*), from Alexandria, Egyptian Mediterranean coast (Said et al. 2006). They ranged from undetectable to 16.0, 44.0–82.0, 10.0–43.0, and 131–240 ng g^{-1}, respectively. Fish from the Suez Gulf in Egypt (*S. rivulatus*, *L. ramada*, *T. puta*, *B.* punctatus, *O. melanura*, and *G. oyena*) were tested for TBT, DBT, and DPhT contamination and their possible detrimental health consequences by Shreadah et al. (2008). The most common component in *O. melanura* was TBT, with 1.17×10^3 ng g^{-1}. TBT's extensive application as biocides in antifouling paints and agriculture, combined with its low metabolism rate, could be the reason of this. The HQ value in this investigation was calculated to be 2.79. In *Liza ramada* sp., the total of Et_4Pb, Me_4Pb, Et_3Pb^+, and Me_3Pb^+ was also examined; it was shown to be dominating, with 276.23 ng g^{-1}. Furthermore, the maximum levels of total Sn and Pb in *L. ramada* and *S. rivulatus* were found to be 0.80 and 7.95 μg g^{-1}, respectively. The levels of Pb were higher than WHO standards, indicating industrial activity in the area. However, the Sn concentrations in the various fish species *under* study were below the WHO-recorded permissible levels. Ashraf et al. (2017) determined OTCs in different fish species from the Arabian Gulf. *Epinephelus tauvina* had the maximum TBT with 98.5 ng g^{-1}, while *Acanthopagrus bifasciatus* had the minimum with 43.7 ngg^{-1}. *Lethrinus miniatus* had the highest levels of TPT with 108 ng g^{-1}, while *Acanthopagrus bifasciatus* had the minimum value of 64.9 ng g^{-1}. *Emperors (Lethrinus miniatus)* had the maximum BTCs with 228.4 ng g^{-1}, while *Acanthopagrus bifasciatus* had a minimum value of 126.4 ng g^{-1}. *Epinephelus tauvina* had the maximum TPhT at 281.9 ngg^{-1}, followed by *Acanthopagrus bifasciatus* at 282 ng g^{-1}. Additionally, the local population's intake of these fish leads to an evaluation of their EDI. It was discovered that *E. microdan* had the highest EDI, 0.01 ng g^{-1}. Bandara et al. (2021) recorded the residual level of TBT contamination in Sri Lanka's biota. TBT in *P. viridis* was discovered to vary in ascending order of body weight from 4.00 ± 1.20 to 42.0 ± 2.20 ng kg^{-1} wet wt. The Dikkowita fishery harbor had the highest TBT level in *P. viridis* (42.0 ± 2.20 ng kg^{-1}). There is a positive association ($p < 0.05$) between the number of male *P. viridis* and TBT levels, which could suggest that aquatic species exposed to high TBT concentrations regularly suffer reproductive damage. Metelkova et al. (2022) measured six species of OTCs in bivalve molluscs (*C. ciliatum*, *M. calcarea*, *C. islandica*), and macrophyte algae (*S. latissima*, *P. palmata*, *U. obscura*, *F. serratus*, *F. distichus*). OTCs include tricyclohexyl tin, tetrabutyltin, DBT, TBT, MBT, and TPhT. The results showed that the examined samples had a considerable amount of OT contamination. In the bivalves and algae samples, the total content of OTC ranged from 13.0–108 and 2.90–75.0 ng g^{-1} (dry wt), respectively, and did not exceed the internationally set limits. The analysis of BT derivatives' degradation indices revealed that TBT and tetrabutyltin were actively transformed in macrophyte algae and that these compounds accumulated in mollusks' soft tissues. The amounts of OTs found in mollusks examined in this investigation were similar to those found in other Arctic regions. It is possible to immobilize or mobilize trace metals from solid phases. Soils, sediments, and suspended particles are a few examples. Direct precipitation/dissolution reactions on organic matters are how they interact with solid solutions. Solid phase metals can be connected to organic matter, OH^-, CO_3^-, or sulfide minerals (in crystalline or amorphous forms), and surface functional groups (Meyer 2002). Thus, biota can alter the trace metals cycle by absorbing them by surface adsorption or assimilation and releasing them through methylation, desorption, or mineralization processes, as illustrated in Fig. 1.7. Aquatic animals can absorb essential metals and consequently also toxic ones, through two different routes: direct absorption from the water and uptake through food. Environmental and biological factors substantially alter the relative relevance of each of these pathways.

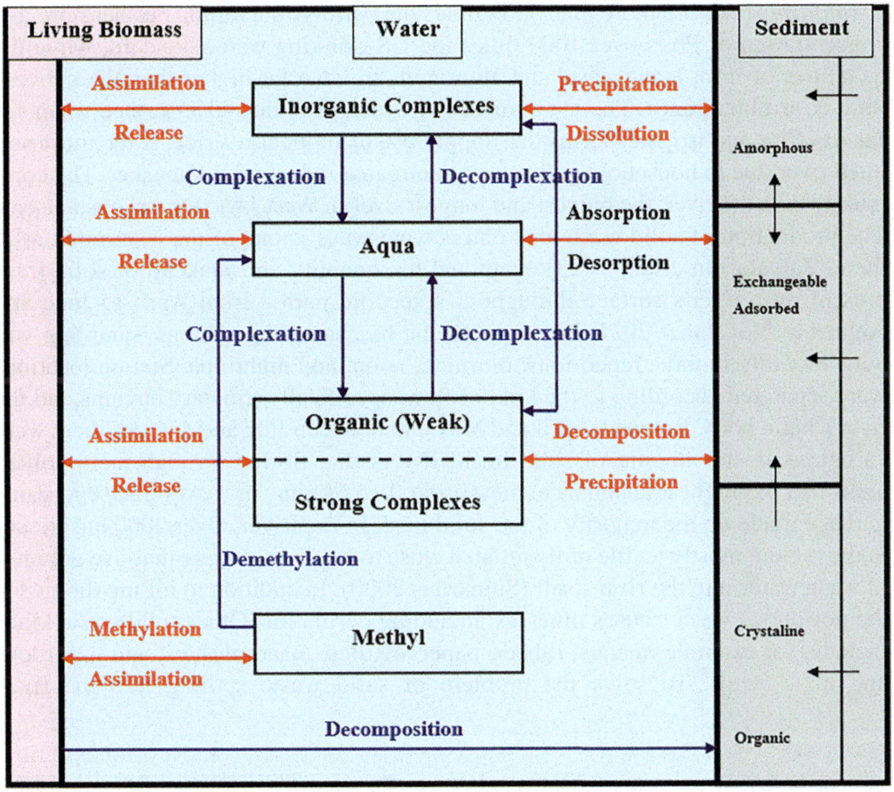

Fig. 1.7 Heavy metals transfers in aquatic environments (Meyer 2002)

1.4.4 Wastewater

The primary sources of direct waste discharge are estuaries, coastal towns, and coastal industries. Most ports have historically developed on estuaries, where they became hubs for commerce and population. Before human awareness of the serious effects of marine pollution, waste from cities and other sources was dumped into estuaries untreated (Bodin et al. 2004). However, some coastal industries, like mariculture installations, were created in response to demands for greater attention and adequate supplies. These industries contributed to some of the direct input of organic pollutants and leftover food into coastal waters.

OMCs have caused significant concern in recent years due to their high toxicity to nontarget organisms even at low concentrations when leached into various environmental compartments (Marques et al. 2016; Gao et al. 2017). The Kahayan River in Kalim antan and the Citarum River in Java were named among the most polluted areas with Chernobyl during 2013 (Asian Development Bank and the World Bank 2013; Asian Development Bank (ADB) 2014). In West Java, Indonesia, where

F. coliform levels are more than 5000 times the permissible limits. Furthermore, the concentration of Pb is over 1000 times the US drinking water standard, while the quantities of iron, manganese, and aluminum are also far higher than the national limit (Carrubba 2020). The Citarum River, which provides 80% surface water for Jakarta City and irrigates farms that supply 5% of Indonesia's rice, is the most polluted river due to household garbage and untreated industrial wastewater. The community uses the river, the biggest and longest river in West Java, for various reasons. The production of solid waste that enters waterways is one of the issues that arise there. Zainalarifin et al. (2023) examined the quantity and makeup of solid trash present on a river's surface throughout a specific period from April to June and August to November 2021 (Fig. 1.8). At the five sampling stations, sampling was done monthly in three repetitions: morning, noon, and nighttime. Station locations were separated according to the kind of flow, specifically tributary streams and the river's main body. Between April and May and between July and August, there were variations in the amount of solid trash. Rivers that flowed through metropolitan areas had the highest abundance, measuring 0.157 kg m^{-3} h^{-1}. At 62.85%, organic garbage made up the majority of the solid trash in the River. Over 2000 businesses in the region, mostly textile mills situated close to the river, release massive volumes of wastewater into the river itself (Sumartias 2020). In addition to killing the rivers, the poisonous waste causes illnesses, including dermatitis. Other solid waste kinds include, for example, metals, rubber, paper, textiles, macroplastics, and hazardous and toxic items. To solve the problem of solid waste spilling into the river,

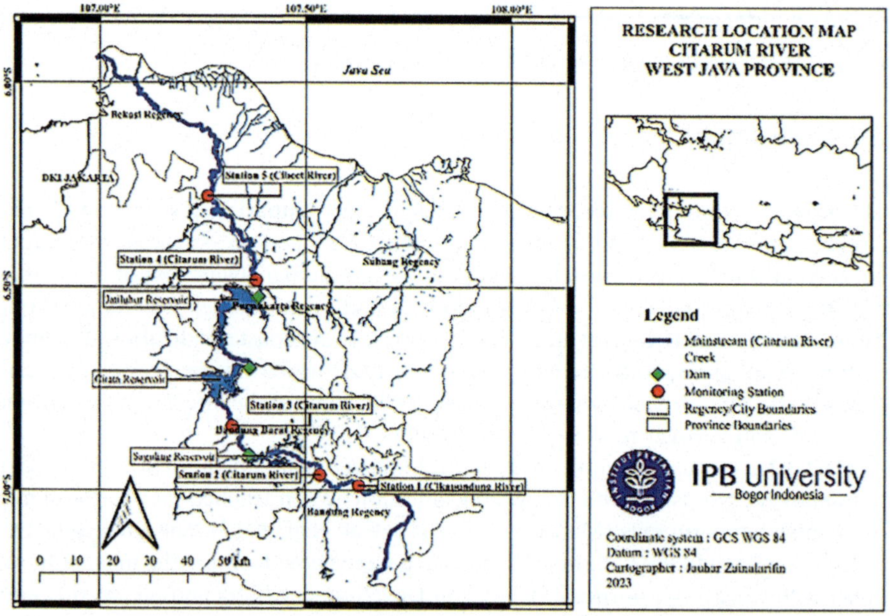

Fig. 1.8 Research location map at Citarum River, West Java Province (Zainalarifin et al. 2023)

communities and local stakeholders must be involved in solid waste management based on sustainability and area.

Bandara et al. (2021) improved and modified the procedure for determining the levels of TBT contamination in Sri Lanka's sediment and coastal waters. TBT concentrations ranged from 303 ± 7.40 to 25.0 ± 4.20 ng L^{-1} in coastal water, while the sediment concentrations ranged from 107 ± 4.1 ng kg^{-1} to 17.0 ± 1.40 ng kg^{-1}.

References

Abbott PJ (1987) Methylcyclopentadienyl manganese tricarbonyl (MMT) in petrol: the toxicological issues. Sci Total Environ 67(2-3):247–253

Abbott JKC, Dougan BA, Xue ZL (2011) Chapter 13: synthesis of organometallic compounds. In: Modern inorganic synthetic chemistry. Elsevier, Amsterdam, pp 269–293. https://doi.org/10.1016/B978-0-444-53599-3.10013-7

ACFA (2010) Africa fuel quality update—progress & challenges. Asian Clean Fuels Assoc 8(5):1–6

Adhikari S, Kaminsky W, Rao Kollipara M (2017) Investigation of the coordination chemistry of multidentate azine Schiff-base ligands towards d6 half-sandwich metal complexes. J Organomet Chem 848:95–103

Allardyce CS, Dorcier A, Scolaro C, Dyson PJ (2005) Development of organometallic (organotransition metal) pharmaceuticals. Appl Organomet Chem 19:1–10

Al-Yousuf MH, El-Shahawi MS, Al-Ghais SM (2000) Trace metals in liver, skin and muscle of Lethrinus lentjan fish species in relation to body length and sex. Sci Total Environ 256(2–3):87–94

Ambasht RS, Ambasht PK (2001) Environment and pollution.3rd edn. CBS Publishers and Distribution, Delhi

Amina LRS, Mandal R, Nouri MH, Murimboh J, Chakrabart CL, Back MH, Gregoire DC, Schroeder WH (1999) Effect of metal/fulvic acid mole ratios on the binding of Ni(II), Pb(II), Cu(II), Cd(II), and Al(III) by two well-characterized fulvic acids in aqueous model solutions. Anal Chim Acta 402:211–221

Andrewes P, Cullen WR, Polishchuk E, Reimer KJ (2001) Antimony biomethylation by the wood rotting fungus Phaeolus Schweinitzii. Appl Organometal Chem 6:473–480

Aran D, Gury M, Jeanroy E (2001) Organo-metallic complexes in an andosol: a comparative study with a Cambisol and Podzol. Geoderma 97(1-2):65–79

Ashraf MW, Salam A, Mian A (2017) Levels of organotin compounds in selected fish species from the Arabian Gulf. Bull Environ Contam Toxicol 98:811–816. https://doi.org/10.1007/s00128-017-2083-9

Asian Development Bank (ADB) (2014) Cleaning up Indonesia's Citarum basin. https://www.adb.org/features/cleaning-indonesias-citarum-basin. Accessed 6 February 2022

Asian Development Bank and the World Bank (2013) Downstream impacts of water pollution in the upper Citarum River, West Java, Indonesia, Water, and Sanitation Program: Technical Paper

Atwood JD (1985) Mechanisms of inorganic and organometal/ic reactions. Brooks/Cole Publishing Company, Pacific Grove

Bandara KRV, Chinthakac SDM, Yasawardened SG, Managea PM (2021) Modified, optimized method of determination of tributyltin (TBT) contamination in coastal water, sediment and biota in Sri Lanka. Mar Pollut Bull 166:112202

Bhatt V (2016) Chapter 7: basic organometallic chemistry. In: Essentials of coordination chemistry. A simplified approach with 3D visuals. Academic, New York, pp 173–190. https://doi.org/10.1016/B978-0-12-803895-6.00007-0

Biester H, Muller G, Scholer HF (2002) Binding and mobility of mercury in soils contaminated by emissions from chloral Kili plants. Sci Total Environ 284:191–203

Birchenough AC, Barnes N, Evans SM, Hinz H, Kronke I, Moss C (2002) A review and assessment of tributyltin contamination in the North Sea, based on surveys of tributyltin tissue burdens and imposex/intersex in the four species of neogastropods. Mar Pollut Bull 44:534–543

Blakemore JD (2022) Electrochemistry in organometallic chemistry. Elsevier, Amsterdam, pp 249–283. https://doi.org/10.1016/b978-0-12-820206-7.00134-7

Blakey DH, Bayley JM, Douglas GR (1992) Induction of chromosomal aberrations in Chinese hamster ovary cells by triethyllead acetate. Mut Res Genet Toxicol 298(1):1–7

Bodin N, Burgeot T, Stanisiere JY, Bocquene G, Menard D, Minier C, Boutet I, Amat A, Cherel Y, Budzinski H (2004) Seasonal variations of a battery of biomarkers and physiological indices for the mussel Mytilus galloprovincialis transplanted into the Northwest Mediterranean Sea. Comp Biochem Physiol Part C 138:411–427

Boening DW (2000) Ecological effects, transport, and fate of mercury: a general review. Chem 40:1335–1351

Briant N, Chouvelon T, Martinez L, Brach-Papa C, Chiffoleau JF, Savoye N, Sonke J, Knoery J (2017) Spatial and temporal distribution of mercury and methylmercury in bivalves from the French coastline. Mar Pollut Bull 114(2):1096–1102

Buck KN, Bruland KW (2005) Copper speciation in San Francisco Bay: a novel approach using multiple analytical windows. Mar Chem 96(1–2):185–198

Buesseler KO, Steinberg DK et al (2000) A comparison of the quantity and composition of material caught in a neutrally buoyant versus surface-tetilered sediment trap. Deep-Sea Res 47:277–294

Caballero B, Olguin N, Campos F, Farina M, Ballester F, Lopez-Espinosa MJ, Llop S, Rodríguez-Farré E, Suñol C (2017) Methylmercury-induced developmental toxicity is associated with oxidative stress and cofilin phosphorylation. Cellular and human studies. NeuroToxicology 59:197–209. https://doi.org/10.1016/j.neuro.2016.05.018

Cances B, Pontheiu M, Castrec-Rouelle M, Aubry E, Benedetti MF (2003) Metal ions speciation in a soil and its solution: experimental data and model results. Geoderma 113(3–4):341–355

Cao J, Lam KC, Dawson RW, Liu WX, Tao S (2004) The effect of pH, ion strength and reactant content on the complexation of Cu^{2+} by various natural organic ligands from water and soil in Hong Kong. Chemosphere 54(4):507–514

Carrubba A (2020) Rotten river: life on one of the world's most polluted waterways–photo essay. https://www.theguardian.com/global-development/2020/nov/02/rotten-river-life-on-one-of-the-worlds-most-polluted-waterways-photoessay. Accessed 26 January 2022

Challa YR, Astudillo LRD, Ramirez A, Escalona A, Martinez G (2008) Distribution of total and organic mercury in superficial soils in the upon Manzanares River watershed, Sucre state, Venezuela. Air Soil Water Res 1:21–29

Chau YK, Wong PTS (1991) Recent developments in speciation and determination of organometallic compounds in environmental samples. Fresenius J Anal Chem 339:640–645

Chen T, Quettee SS, Ambrose RF (2001) Trace metals in fish and invertebrates of three Califorina costal wetlands. Mar Pollut Bull 42(3):224–232

Chen M, Tian X, Li W, Bekyarova E, Li G, Matthew L, Moser R, Haddon C (2016) Application of organometallic chemistry to the electrical interconnection of graphene nanoplatelets. Chem Mater 28(7):2266. https://doi.org/10.1021/ACS.CHEMMATER.6B00217

Cheung KC, Poon BHT, Lan CY, Wong MH (2003) Assessment of metal and nutrient concentrations in river water and sediment collected from the cities in the Pearl River Delta, South China. Chemosphere 52:1431–1440

Christodoulou CV, Ferry DR, Fyfe DW, Young A, Doran J, Sheehan TMT, Eliopoulos A, Hale K, Baumgart J, Sass G, Kerr DJ, Clin J (1998) Anti-proliferative activity and mechanism of action of titanocene dichloride. Br J Cancer 77(12):2088–2097

Cini M, Bradshaw TD, Lewis W, Woodward S (2013) Structure and reactivity of a preactivated sp2–sp3 diboron reagent: catalytic regioselective boration of α,β-unsaturated conjugated compounds. Eur J Org Chem 76:3997–4007

Clarckson TW, Magos L (2006) The toxicology of mercury and its chemical components. Crit Rev Toxicol 36:609–662

Cole RF, Mills GA, Parker R, Bolam T, Birchenough A, Kroger S, Fones GR (2015) Trends in the analysis and monitoring of organotins in the aquatic environment. Trends Environ Anal Chem 8:1–11

Crabtree RH (2009) The organometallic chemistry of the transition metals. Wiley, Hoboken

Craig PJ (1986) Organometallic compounds in the environment. Longman, London

Craig PJ (2003) Organometallic compounds in the environment.2nd edn. Wiley, Chichester

Craig PJ (2023) Organometallic compounds in the environment.2nd edn. Wiley, Chichester, p 434

Craig PJ, Glocking F (1988) The biological alkylation of heavy elements. The Royal Society of Chemistry, London

Craig PJ, Jenkins RO (2004) Organometallic compounds in the environment: an overview. In: Hirner AV, Emons H (eds) Organic metal and metalloid. Springer, Cham

Craig PJ, Miller D (1997) Analysis and speciation of organometallic compounds in the marine environment. In: Gianguzza A, Pelizzetti E, Sammartano S (eds) Marine chemistry - an environmental analytical chemistry approach, vol 25. Kluwer Academic, Dordrecht, pp 161–172

Craig PJ, Eng G, Jenkins RO (2003) Occurrence and pathways of organometallic compounds in the environment—general considerations. In: Craig PJ (ed) Organometallic compounds occurrence in the environment2nd edn. Wiley, Chichester, pp 1–56

Datta MG (2023) Pollution in marine ecosystem: impact and prevention. In: Soni R, Suyal DC, Morales-Oyervides L, Fouillaud M (eds) Current status of marine water microbiology. Springer, Singapore. https://doi.org/10.1007/978-981-99-5022-5_8

De Stefano C, Foti C, Gianguzza A, Sammartano S (2000) Speciation of low molecular weight ligands in natural fluids: protonation constants and association of open chain polyamines with the major components of seawater. Anal Chim Acta 418:43–51

Ding Q, Cheng G, Wang Y, Zhuang D (2017) Effects of natural factors on the spatial distribution of heavy metals in soils surrounding mining regions. Sci Total Environ 578:577–585

Egorova KS, Ananikov VP (2017) Toxicity of metal compounds: knowledge and myths. Organometallics 36(21):4071–4090

Farkas A, Salanki J, Specziar A (2003) Age and size-specific patterns of heavy metals in the organs of freshwater fish Abramis brama L. populating a low-contaminated site. Water Res 37:959–964

Feng X, Zhang K, Hempenius MA, Vancso GJ (2015) Organometallic polymers for electrode decoration in sensing applications. RSC Adv 5(129):106355–106376. https://doi.org/10.1039/C5RA21256A

Filgueiras AV, Lavilla I, Bendicho C (2002) Chemical sequential extraction for metal partitioning in environmental solid samples. J Environ Monit 4:823–857

Florence TM (1982) The speciation of trace elements in waters. Talanta 29(5):345–364

Gafurov ZN, Kantyukov AO, Kagilev AA, Sinyashin OG, Yakhvarov DG (2021) Electrochemical methods for synthesis and in situ generation of organometallic compounds. Coord Chem Rev 442:213986

Gaiero DM, Probst JL, Depetris PJ, Lelyter L, Kempe S (2002) Riverine transfer of heavy metals from Pantagonia to southwestern Atlantic Ocean. Reg Environ Chang 3(1–3):51–64

Gallert C, Winter J (2002) Bioremediation of soil contaminated with alkyllead compounds. Water Res 36(12):3130–3140

Gao JM, Wu L, Chen YP, Zhou B, Guo JS, Zhang K, Ouyang WJ (2017) Spatiotemporal distribution and risk assessment of organotins in the surface water of the three gorges reservoir region, China. Chemosphere 171:405–414

Gasser G, Ott I, Metzler-Nolte N (2011) Organometallic anticancer compounds. J Med Chem 54(1):3–25. https://doi.org/10.1021/jm100020w

Gianguzza A, Pelizzetti E, Sammartano S (2002) Chemistry of marine water and sediments. Springer, Berlin. https://doi.org/10.1007/978-3-662-04935-8

Gordon AS, Donat JR, Kango RA, Dyer BJ, Stuart LM (2000) Dissolved copper-complexing ligands in cultures of marine bacteria and estuarine water. Mar Chem 70:149–160

Guven K, Ozbay C, Unlu E, Satar A (1999) Acute lethal toxicity and accumulation of copper in Gammarus pules (L.) (Amphipoda). Turk J Biol 23:51–56

Harrison RM, Rapsomanikis S (1989) Environmental analysis using chromatography interfaced with atomic spectroscopy. Wiley, New York

Hartinger CG, Metzler-Nolte N, Dyson PJ (2012) Challenges and opportunities in the development of organometallic anticancer drugs. Organometallics 31:5677–5685

Henriques B, Rodrigues SM, Coelho C, Cruz N, Duarte AC, Romkens PFAM, Pereira E (2013) Risks associated with the transfer of toxic organo-metallic mercury from soils into the terrestrial feed chain. Environ Int 59:408–417

House JE, House KA (2016) Descriptive inorganic chemistry.3rd edn. Academic, New York, pp 371–393. https://doi.org/10.1016/B978-0-12-804697-5.00022-1

Hu H, Li Z, Feng Y, Liu Y, Xue J, Davis M, Liang Y (2016) Prediction model for mercury transfer from soil to corn grain and its cross-species extrapolation. J Integr Agric 15(10):2393–2402

Jacobsen JA, Asmund G (2000) TBT in marine sediments and blue mussels (Mytilus edulis) from Central-West Greenland. Sci Total Environ 245:131–136

Jacobsen JA, Strand J (2000) Depositional profiles of Butyltin residues in marine sediments from the sound between Sweden and Denmark. Poster session presented at SETAC 21st annual meeting, Nashville, United States

Jaishankar M, Tseten T, Anbalagan N, Mathew BB, Beeregowda KN (2014) Toxicity, mechanism, and health effects of some heavy metals. Interdiscip Toxicol 7(2):60–72

Jantzen E, Prange A (1995) Organometallic species of the elements tin, mercury and lead in sediments of the longitudinal profile of the river Elbe. Fresenius J Anal Chem 353:28–33

Jaouen G (2006) Bioorganometallics. Wiley-VCH, Weinheim

Jaouen G, Dyson P (2007) Comprehensive organometallic chemistry III, vol 12. Elsevier, Oxford, p 445

Joseph NK (2002) Heavy metal distribution and enrichment at port-Reitz Creek, Mombasa, (western Indian Ocean). Mar Sci 1(1):65–70

Kimbrough KL, Johnson WE, Lauenstein GG, Christensen JD, Apeti DA (2008) An assessment of two decades of contaminant monitoring in the nation's coastal zone. NOAA Technical Memorandum NOS NCCOS, Silver Spring, p 105

Köpf-Maier P (1990) Intracellular localization of titanium within xenografted sensitive human tumors after treatment with the antitumor agent titanocene dichloride. J Struct Biol 105(1-3):35–45. https://doi.org/10.1016/1047-8477(90)90096-u

Köpf-Maier P (1999) Antitumor activity of titanocene dichloride in xenografted human renal-cell carcinoma. Anticancer Res 19:493–504

Köpf-Maier P, Kopf H (1994) Metal compounds in cancer therapy. Springer, Dordrecht, pp 109–146

Kumar A, Avula B, Yadav S, Singh Y, Eraganaboyina S, Zyryanov GV, Nemallapudi BR, Singh AK, Bhimireddi R (2023) Synthesis, properties, environmental stability and practical application of organometallic compounds: a comprehensive review. Inorg Chem Commun 158(2):111567. https://doi.org/10.1016/j.inoche.2023.111567

Lagerström M, Strand J, Eklund B, Ytreberg E (2017) Total tin and organotin speciation in historic layers of antifouling paint on leisure boat hulls. Environ Pollut 220(Part B):1333–1341. https://doi.org/10.1016/j.envpol.2016.11.001

Leadbeater NE (2007) Comprehensive organometallic chemistry III. In: Polymer-supported organometallic catalysts. Elsevier, Amsterdam, pp 663–753. https://doi.org/10.1016/b0-08-045047-4/00179-5

Lee R, Oshima Y (1998) Effects of selected pesticides, metals and organometallics on development of blue crab (Callinectes sapidus) embryos. Mar Environ Res 46(1–5):479–482

Lindqvist O, Jernelov A, Johansson K, Rodhe H (1984) Mercury in the Swedish environment: global and local sources. In: Report of the workshop. Swedish Environmental Protection Board, Solna, pp 105–110

Lippert B (1999) Cisplatin-chemistry and biochemistry of a leading anticancer drug. Verlag Helvetica Chimica Acta, Zürich, pp 1–584. https://doi.org/10.1002/9783906390420

Lobinski R, Dirks WR, Szpunar-Lobinska J, Adams FC (1994) Anal Chim Acta 286:381
Loyaux-Lawniczak S, Lecomte P, Ehrhardt J-J (2001) Behaviour of heaxavalent chromium in apolluted groundwater: redox processes and immobilization in soils. Environ Sci Technol 35(7):1350–1357
Lu M, Wu XJ, Zheng DC, Liao Y (2012) Distribution of PCDD/Fs and organometallic compounds in sewage sludge of wastewater treatment plants in China. Environ Pollut 171:78–84
Malle KG (1992) Zinc in the environment. Acta Hydrchim Hydrobiol 20:196–204
Marques RC, Abreu L, Bernardi JVE, Dorea JG (2016) Neurodevelopment of Amazonian children exposed to ethylmercury (from thimerosal in vaccines) and methylmercury (from fish). Environmental Research 149:259–265
McCoy CP, Hara TM, Bennett LW, Boyle CR, Lynn BC (1995) Liver and kidney concentrations of zinc, copper and cadmium in channel catfish (Ictalurus punctatus): variation due to size, season and health status. Vet Hum Toxicol 37(1):11–15
Meanwell NA (2018) Fluorine and fluorinated motifs in the design and application of bioisosteres for drug design. J Med Chem 61(14):5822–5880
Melchart M, Sadler PJ (2005) Bioorganometallics. WileyVCH, Hoboken, p 39
Metelkova L, Zhakovskaya Z, Kukhareva G et al (2022) Organotin compounds (OTs) in surface sediments, bivalves and algae from the Russian coast of the Barents Sea (Kola peninsula) and the Fram Strait (Svalbard archipelago). Environ Sci Pollut Res 29:34659–34669. https://doi.org/10.1007/s11356-021-18091-0
Meyer J (2002) The utility of the terms "bioavailability" and "bioavailable fraction" for metals. Mar Environ Res 53:417–423
Michalke K, Wickenheiser A, Mehring A, Hirner AV, Hense R (2000) Production of volatile derivatives of metal(loid)s by microflora involved in anaerobic digestion of sewage sludge. Appl Environ Microbiol 66:2791–2796
Morillo J, Usero J, Gracia I (2002) Partitioning of metals in sediments from the Odiel river (Spain). Environ Int 28:263–271
Morita M, Yoshinaga J, Edmonds JS (1998) The determination of mercury species in environment and biological samples (technical report). In: Pure and applied chemistry1st edn. IUPAC, London, pp 1585–1615
Morse JW, Luther GW (1999) Chemical influences on trace metal-sulfide interactions in anoxic sediments. Geochim Cosmochim Acta 63:3373–3379
Nakanishi T (2008) Endocrine disruption induced by organotin compounds; organotins function as a powerful agonist for nuclear receptors rather than an aromatase inhibitor. J Toxicol Sci 33:269–276
Ndagi U, Mhlongo N, Soliman ME (2017) Metal complexes in cancer therapy - an update from drug design perspective. Drug Dev Ther 11:599–616
Nierop KGJ, Jansen B, Vrugt JA, Verstraten JM (2002) Copper complexation by dissolved organic matter and uncertainty assessment of their, stability constants. Chemosphere 49(10):1191–1200
Nriagu JO (1980) Nickel in the environment. Wiley, Chichester
Nriagu JO (1994) Arsenic in the environment: part I: cycling and characterization. Wiley, New York
Nussey G, Van Vuren JHJ, du Preez HH (2000) Bioaccumulation of chromium, manganese, nickel and lead in the tissues of the moggel, Labeo umbratus (Cyprinidae), from Witbank dam, Mpumalanga. Water SA 26:269–284
Palmer SL, Presley BJ (1993) Mercury bioaccumulation by shrimp (Penaeus azlecus) transplanted to Lavaca Bay, Texas. Mar Pollut Bull 26:264–266
Pan LB, Ma J, Wang XL, Hou H (2016) Heavy metals in soils from a typical county in Shanxi Province, China: levels, sources and spatial distribution. Chemosphere 148:248–254
Parham S, Sarvari H, Parham S-S, Nur H (2023) Emerging role of organometallic compounds for drug delivery, renewable energy, and wastewater treatment. Wiley, New York. https://doi.org/10.1002/9783527840946.ch17
Pearson R (1968a) Hard and soft acids and bases, HSAB, part I, fundamental principles. J Chem Educ 45:581–587

Pearson R (1968b) Hard and soft acids and bases, HSAB, part II, underlying theories. J Chem Educ 45:643–648
Perez PJ (2024) Adv Organomet Chem 82:1–286. https://doi.org/10.1016/S0065-3055(24)00030-3
Pourang N, Amini G (2001) Distribution of trace elements in tissues of two shrimp species from Persia Gulf and effects of storage temperature on elements transportation. Water Air Soil Pollut 129:229–243
Powell P (1988) Principles of organometallic chemistry.2nd edn. Springer, Cham
Pyrzynska K (1996) Organolead speciation in environmental samples: a review. Microchim Acta 122(3):279–293
Quaghebeur M, Rate A, Rengel Z, Hinz C (2005) Heavy metals in the environment desorption kinetics of arsenate from kaolinite as influenced by pH. J Environ Qual 34:479–489
Rasmussen SC (2015) The 18-electron rule and electron counting in transition metal compounds: theory and application. ChemTexts 1:10. https://doi.org/10.1007/s40828-015-0010-4
Reis AT, Rodriques SM, Davidson CM, Pereira E, Duarte AC (2010) Extractability and mobility of mercury from agricultural soils surrounding industrial and mining contaminated areas. Chemosphere 81(11):1369–1377
Said TO, Farag R, Younis A, Shreadah MA (2006) Organotin species in fish and bivalve samples collected from the Egyptian Mediterranean coast of Alexandria. Egypt Bull Environ Contam Toxicol 77:451–458. https://doi.org/10.1007/s00128-006-1086-8
Schaanning M, Naes K, Egeberg PK, Rome F (1988) Cycling of manganese in permanently anoxic Drammens fjord. Mar Chem 23:365–382
Schfer S, Ott I, Gust R, Sheldrick WS (2007) Influence of the polypyridyl (pp) ligand size on the DNA binding properties, cytotoxicity and cellular uptake of organoruthenium(II) complexes of the type $[(\eta^6\text{-}C_6Me_6)Ru(L)(pp)]^{n+}$ [L = cl, n = 1; L = $(NH_2)_2CS$, n = 2]. Eur J Inorg Chem 2007:3034–3046
Sekhar CKC, Chary NS, Kamala T, Raj DSS, Rao AS (2003) Fractionation studies and bioaccumulation of sediment-bound heavy metals in Kolleru lake by edible fish. Environ Int 29:1001–1008
Selin NE (2009) Global biogeochemical cycling of mercury: a review. Annu Rev Environ Resour 34:43–63
Ser PH, Omi S, Shimizu-Furusawa H, Yasutake A, Sakamoto M, Hachiya N et al (2017) Differences in the responses of three plasma selenium-containing proteins in relation to methylmercury-exposure through consumption of fish/whales. Toxicol Lett 267:53–58
Shekhar S, Khan AM, Sharma S et al (2022) Schiff base metallodrugs in antimicrobial and anticancer chemotherapy applications: a comprehensive review. Emerg Mater 5:279–293. https://doi.org/10.1007/s42247-021-00234-1
Shreadah MA, Said TO, Abd El Ghani SA, Ahmed AM (2008) Alkyllead and alkyltin species in different fishes collected from the Suez gulf. Egypt Egypt J Aquat Res 34:64–73
Siegl H, Sigel A (1993) Metal ions in biological systems, vol. 29: biological properties of metal alkyl derivatives. Dekker, New York
Siegler WR, Nierenberg DW, Hickey WF (1999) Fatal poisoning from liquid dimethylmercury: a neuropathologic study. Hum Pathol 30(6):720–723
Sigel H, Sigel A (1984) Circulation of metals in the environment. Marcil Dekker, Basel
Skeaff J, King P (2000) Reaction kinetics of nickel metal in aqueous media at pH 6 for Hazard indetification. Report MMSL 99-067 (CR). Final Report, Cannet, Mining and Mineral Sciences Laboratories, Natural Resources Canada
Storelli MM, Marcootrigiano GO (2002) Mercury speciation and relationship between mercury and selenium in liver of Galeus melastomus from the Mediterranean Sea. Bull Environ Contamin Toxicol 69:516–522
Strohfeldt K, Tacke M (2008) Bioorganometallic fulvene-derived titanocene anti-cancer drugs. Chem Soc Rev 37:1174–1187
Sumartias S (2020) The community responses over the Citarum river West Java. Eur J Mol Clin Med 7(1):3724–3732
USEPA (2007) Framework for metal risk assessment. USEPA, Washington

Wang D, Lippard SJ (2005) Cellular processing of platinum anticancer drugs. Nat Rev Drug Discov 4(4):307–320. https://doi.org/10.1038/nrd1691

Wu SH, Shi YX, Zhou SL, Wang CH, Chen H (2016) Modelling and mapping of critical loads for heavy metals in Kunshan soil. Sci Total Environ 569:191–200

Yan YK, Melchart M, Habtematiam A, Sadler PJ (2005) Organometallic chemistry, biology and medicine: ruthenium arene anticancer complexes. Chem Commun 2005:4764–4776

Zainalarifin J, Effendi H, Taryono (2023) Abundance and composition of solid waste in the Citarum River, West Java Provinc. IOP Conf Ser Earth Environ Sci 1266:012056

Zaki M, Hairatb S, Aazama ES (2019) Scope of organometallic compounds based on transition metal-arene systems as anticancer agents: starting from the classical paradigm to targeting multiple strategies. RSC Adv 9:3239–3278

Zhaou YJ, He JC, Chen Q, He J, Hou HQ, Zheng Z (2011) Evaluation of 206 nm UV radiation transformation for degrading organometallics in wastewater. Chem Eng J 167(1):22–27

Zimmermann LT, Santos DB, Naime AA, Leal RB, Dorea JG, Barbosa Jr F et al (2013) Comparative study on methyl- and ethylmercury-induced toxicity in C6 glioma cells and the potential role of LAT-1 in mediating mercurial-thiol complexes uptake. Neurotoxicology 38:1–8

Zuniga MC, Jover E, Arancibia V, Bayona JM (2009) Development of a methodology for the simultaneous determination of inorganic and organolead compounds using supercritical fluid extraction followed by gas chromatography-mass spectrometry and its application to environmental matrices. Talanta 80(2):504–510

Zuo Z, Chen S, Wu T, Zhang J, Su Y, Chen Y, Wang C (2011) Tributyltin causes obesity and hepatic steatosis in male mice. Environ Toxicol 26(1):79–85

Chapter 2
Extraction Methodology of OMCs [Stability, Evaluation of the Extraction Recoveries, Improving the Reliability of Speciation Analysis]

Abstract Extracting OMCs from environmental samples is extremely complex because of two competing factors: achieving sufficient recovery and preventing losses, particularly the degradation of the chemicals or compounds. The recoveries and the assessment of extraction techniques for OMCs are documented in this chapter. A wide advances in analytical methods and all the latest discoveries in this field are discussed. This chapter intends to summarize the most recent approaches and techniques discovered for improving the reliability of speciation analysis of OMCs. Numerous environmental and biological studies recommend derivatization procedures for the determination of OTCs. A focus on hydride generation, alkylation by GRs, and ethylation with sodium tetraethyl borates as derivatization procedures is reported. For measurements, several hyphenated approaches based on GC-AAS, MIP-AES, and GC-MIP-AES have been used. A multifaceted analytical approach to identify OMC species, with a focus on Sn, Hg, and Pb stability as a governing factor, is highlighted. It is necessary to assess recovery from CRMs, recovery from the isotope dilution methodology, and the standard addition method to obtain valuable information.

Keywords Sample treatment techniques · Stability · Speciation · GC · HPLC · Atomic absorption · Recovery

2.1 Preface

Numerous studies were recorded in various shipping regions (Castro et al. 2018; Metelkova et al. 2022; Zhakovskaya et al. 2022; Qi et al. 2024). In AAS, the majority of instruments utilized in these activities have been readily available, whether by atomic fluorescence, emission, or absorption (Zou et al. 2018; Yahya et al. 2022). These methods ascertain the sample's overall metal content but do not reveal the chemical identity of the material. Nowadays, a deeper understanding of the intricate and crucial cycles of organometallic forms of metals and metalloids in the

T. O. Said, G. M. El Zokm, *Organometallic Compounds in the Marine Environment*, https://doi.org/10.1007/978-3-031-99189-9_2

environment will be facilitated by the combination of atomic spectrometry detectors with gas or liquid separation techniques (Braun 2024). It is interesting to consider the recent history and observe how the progress of the knowledge of trace metal toxicity and cycling is directly influenced by the available equipment (Patle et al. 2022). When excess trace metals are found in the biota, they are usually regarded as toxic. It is now essential to address some of the real difficulties about the existence of chemical forms of the metals rather than the overall amount of metals in the sample, even though this broad framework can be taken as a general guideline. The four kinds of OMCs—precursor technique, hydrometallation, organic halides, and metathesis—as well as their common synthesis process have received special attention. Numerous widely used analytical techniques have been created (Adams and Sofie 2000; Kucharski et al. 2022). The topic of derivatization techniques for identifying OTCs is concerned (Morabito et al. 2002; Chen et al. 2024). They include hydride formation, alkylation by GRs, ethylating by sodium tetraethyl borates, speciation of OHgC, and speciation of OLCs. It is also stated that chemical species are stable in environmental matrices, especially Sn, Hg, and Pb species (Gómez-Ariza et al. 2000; Gworek et al. 2020; Coelho-Junior et al. 2023). Finally, the evaluation of OMCs extraction recoveries from CRMs, isotope dilution approach, standard addition method, and use of internal standards are to be considered. However, there has been significant advancement in speciation analysis in recent years (Craig and Jenkins 2004; Yang et al. 2023). The OMC fragment needs to be isolated from its ambient binding and then measured to perform speciation studies in the environment (Llaver et al. 2021). All the most recent developments in analytical methods are covered in this chapter, which encompasses broad advancements in analytical methods. This review seeks to provide the most recent approaches and strategies created for this reason. Many sample treatment techniques are presented and explained, with an emphasis on the application of cutting-edge nanomaterials and unique solvents in solid phase and liquid-liquid microextractions. Each pertinent element is considered, and the significance of its speciation analysis is emphasized. Furthermore, a comprehensive review of instrumental methods—from chromatographic separations to electrochemical speciation analysis—that are intended to separate and quantify metal and metalloid species is provided. In this book, special emphasis is given to the environmental friendliness of these innovations, taking into account how well they follow the green chemistry idea and critically examining their environmental implications. The procedures ought to be carried out such that they minimize interferences; avoid loss, contamination, or speciation change; and efficiently separate the analyte from the matrix interference. Both the separation and determination in analytical systems are referred to be interfaced or hyphenated. This is particularly significant because, despite having actual negative effects on the environment and/or toxicity, these processes frequently occur in the environment at deficient concentrations of ppb and ppm (Craig et al. 2003). A method that is sensitive enough to detect the metal, OMC fragment, or entire compound, such as Hg,

CH_3Hg^+, or CH_3HgCl, respectively, must be used. The primary techniques for detection include 1) AAS (flame, graphite, Zeeman, hydride generation/quartz furnace) is the primary detection method, (2) AFS (through hydride generation or alone), (3) voltammetry, (4) X-ray neutron methods, and (5) AES (typically ICP). However, the primary methods of separation are (1) GC (conventional or capillary), (2) boiling point-based thermal deposition techniques, (3) HPLC, (4) flow injection techniques, and (5) ion chromatography. OMCs are not volatile enough to pass through GC; hence, they may induce by derivatization. This is usually achieved by (formally) SN_2 through hydridation using $NaBH_4$, ethylation by $NaB(C_2H_5)_4$, or GR group, as shown in Eq. (2.1):

$$As(CH_3)_2X \xrightarrow{NaBH_4} As(CH_3)_2H \tag{2.1}$$

X environmental counter-ion.

Nowadays, it is commonplace to combine these methods of separation and detection, which offers a vast array of analytical tools. It would be difficult to understand OMCs in the environment without these. Through the leaching of the coatings, OMCs reach the aquatic environment and continue to build up in bottom sediments. Several studies have been conducted in different shipping zones (Castro et al. 2018; Metelkova et al. 2022; Zhakovskaya et al. 2022; Qi et al. 2024). Instrumentation used in these endeavors has been widely accessible in atomic spectrometry, whether through atomic fluorescence, emission, or absorption (Zou et al. 2018; Yahya et al. 2022). These methods ascertain the sample's overall metal content but do not reveal the chemical identity of the material. It is interesting to consider the recent history and observe how the progress of the knowledge of trace metal toxicity and cycling is directly influenced by the available equipment (Patle et al. 2022). When excess trace metals are found in the biota, they are usually regarded as toxic. It is now essential to address some of the real concerns regarding the existence of different forms of the metals rather than the overall level of metals in the sample, even though this broad framework can be taken as a general guideline. As per the IUPAC definition, "speciation" refers to the method that provides proof of an analyte's atomic or molecular form. The four kinds of OMCs—precursor technique, hydrometallation, organic halides, and metathesis—as well as their common synthesis process have received special attention; in addition, their stability in different matrices is also declared (Gworek et al. 2020; Coelho-Junior et al. 2023). Since the OMC fragment is typically attached to a complex environmental matrix that may not be recognized, a complete speciation investigation is typically not feasible (Craig and Jenkins 2004; Yang et al. 2023). All the most recent developments in analytical methods are covered in this chapter, which encompasses broad advancements in analytical methods. Many sample treatment techniques are presented and explained, with an emphasis on the application of cutting-edge nanomaterials and unique solvents in solid phase and liquid-liquid microextractions. Each pertinent element is considered, and the significance of its speciation analysis is emphasized. Furthermore, a comprehensive review of instrumental methods from chromatographic separations to electrochemical speciation analysis that are intended to separate and quantify metal and metalloid species is provided.

2.2 Speciation

It is challenging to establish speciation in real samples because element species are typically found at low concentration levels and sample matrix concurrents. To isolate the target from the sample matrix and enhance the target species, sample pretreatments are typically necessary. OMC speciation is a crucial topic because their chemical species have a major impact on elemental toxicity, bioavailability, and mobility in the environment and biosphere (Moreda-Piñeiro et al. 2013). Since most analytical equipment cannot directly handle liquid or solid samples, pretreatment techniques are required to remove the essential components from complex liquid and solid matrices. Preparing an authentic matrix sample for analysis is the main objective of the sample pretreatment procedure. Analyte-matrix interactions are nearly always altered by this procedure. The effectiveness, repeatability, and various procedures are all concerned. They are achieved by the analytes and the matrices' physical and other characteristics. The most costly and time-consuming step in analytical procedures is usually sample preparation. Furthermore, trace analysis frequently calls for enormous sample quantities, which can be very laborious and time-consuming to handle. In the past, it has been believed that the sample preparation phase of the analytical process is the most contaminating. Analytical Green Chemistry (AGC) is becoming more and more popular in analytical labs. According to Armenta et al. (2008), the idea behind AGC, which promotes sustainable growth, is to minimize or completely eradicate the environmental damage caused by analytical processes. The concepts that GC suggested for the analytical process include waste prevention, safer solvents and auxiliaries, energy efficiency, real-time analysis that prevents pollution, and the use of chemistry that is inherently safer that prevents accidents (Anastas and Warner 1998). Consequently, some or all AGC concepts mentioned above are included in several new sample pretreatments. New pretreatment techniques also reduce the amount of time spent on sample preparation by using less organic solvent. Moreover, there are little analyte losses, reduced sample amounts required for analysis, and increased sensitivity when extraction and analysis are done online.Various strategies can be employed to obtain sample pretreatments; some of these procedures are low-tech and easy to use without specialist equipment. Procedures for liquid-liquid extraction and distillation fall under this category. But certain methods can be more complex than others. One such method is pressured liquid extraction, which is expensive to acquire and operate. Additionally, new pretreatment methods downsize sample preparation, which lowers the need for organic solvents. Furthermore, a reduced sample volume for analysis, insignificant analyte losses, and increased sensitivity are the outcomes of linking online extraction and analysis. Sample pretreatments can be accomplished by a variety of methods, some of which are easy to use, low cost, and don't require specialist tools. This pertains to the processes of distillation and liquid-liquid extraction. Some methods, meanwhile, can be more complex. One such method is pressured liquid extraction, which is expensive to acquire and maintain. The objectives of sample pretreatment techniques are the same regardless of complexity:

1. High ion concentrations from groundwater and river waters; high salinity (seawater samples); interferences from minerals or airborne particles; or superior selectivity through the elimination of possible pollutants. Thus, among other things, clearing the matrix components can prevent any degradation from different tools.
2. The amounts of newly discovered pollutants in airborne particulate matter and open seawaters are observed in nanogram levels. Increasing the sample concentration (preconcentration procedures) results in an increase in sensitivity.
3. It could be necessary to transform the analyte into a more appropriate form. This includes metals and organometals derivatizing to organic chelates and/or OMCs derivatizing into volatile species using GR's reaction.
4. The technique of derivatization has the following benefits: (1) more volatile species; (2) more stable species; (3) more detectable complexes; and (4) sensitivity. The disadvantages of derivatization processes include the time, the absence of suitable reagents for the analyte, and the need to get rid of the excess derivatization reagent. The analyte must be moved to a solvent that is appropriate for the analytical method being employed to facilitate the analytical determination.
5. The isolation of certain species, if at all feasible (speciation research). The adoption of chromatographic methods makes this purpose less significant.
6. The development of a strong, dependable procedure that is unaffected by changes in the sample matrix. In recent years, there have been trends toward:

   (a) Using small initial sample sizes; a few sample amounts of milliliters are sufficient.
   (b) Easy-to-use automation techniques that minimize sample and keep clean samples include pressurized liquid extraction and extraction techniques assisted by microwave or ultrasound.
   (c) Miniaturization includes new extraction techniques, as well as miniaturized devices for processes.
   (d) Using non-exhaustive extraction techniques, which only extract or preconcentrate a very small amount of the analyte for further analysis.
   (e) Precise timeframe. High selectivity in the extraction process, ordered mesoporous materials, as well as different solvents.
   (f) Safe pretreatment and extraction time; for example, pressured liquid extraction and microwave-assisted extraction reduce times, 1–5 min.
   (g) A reduction number of extracting cycles and an increase in extraction efficiency without compromising the intended result; for example, liquid-liquid extraction techniques typically require two extraction steps. While pressur-

ized liquid extraction only requires one cycle, Soxhlet extractor cycles typically involve about 24 cycles for the isolation of organic components.

(h) According to high-throughput performance, many samples can be pretreated or examined in a specific amount of time using modern analytical detectors. To pretreat many samples within a reduced time, microwave-assisted extraction methods allow for real-time connection to analytical equipment.

(i) Moving the treated sample to the detection systems is part of the online coupling with analytical tools. Examples of this include the preconcentration of analytes using solid phase micro-extraction linked to GC or the continuous digestion/extraction devices used to identify metals from solid samples.

(j) Decrease in the impact on the environment and enhancement of operator safety for the analytical processes. When it comes to getting rid of or using less organic solvents and/or extremely hazardous or ecotoxic compounds, there are four main priorities: As part of the AGC's guiding principles, the analytical laboratory acknowledged the need to reduce energy use and prevent waste output. The application of techniques that minimize or do away with the use of goods, byproducts, solvents, reagents, and feedstock that are detrimental to the habitat and health. The theoretical underpinnings, experimental benefits, and drawbacks of several techniques, in addition, the use of these methods for OMC preconcentration and separation from environmental and biological specimens, are examined (Dakova et al. 2024).

OMCs with a strong covalent bond are created when specific ingredients come together (Malik et al. 2012). Toxicology is often caused by membrane barrier penetration and bioaccumulation in fatty tissues. Thus, although CH_3Hg^+ is a teratogen and toxin that crosses the placenta and brain, ionic Hg(II) is toxic to mammals and corrosive. Generally, only Sn, Ge, and Hg are notable examples—go through biomethylation. Methyl donors in bioalkylation processes include methylcobalamin and S-adenosylmethionine. Except for Co in vitamin B12, microorganisms in sediments and soils are typically the only ones capable of biomethylating metals (Malik et al. 2012). The OMC fragment must be freed from its environmental binding and measured to facilitate speciation analysis in the environment (Hirner and Emons 2004). The process by which a group within a species develops new characteristics and separates from other members of the same species is known as speciation. The new species will differ from its predecessors due to the traits of the new group or the requirements of a different environment. Over the last 20 years, speciation analysis has grown to be a significant area of analytical chemistry. It is possible to distinguish between various kinds in this field. First, OMC species that are anthropogenically released into the environment (such as alkyl lead and alkyl- and aryl tin) or those (such as mercury, silver, Ge, and selenium) whose metabolism (detoxification) or transport mechanisms (oceanic volatilization) entail forming M-C bond naturally by methylation are important in environmental chemistry. Second, there is interest in figuring out metalloporphyrin, which is important in samples like crude oils and gasoline, which behave differently during refining and burning processes.

Lastly, monitoring metabolites of metal-containing medications is creating a fascinating array of new clinical research uses. Due to historical usage and inappropriate disposal, as well as a failure to assess potential contaminated areas, many of these locations are still utilized today for grazing or agricultural. These days, exposure to toxins like mercury, lead, or tin could happen if their distribution and speciation in soils are not properly evaluated, as well as their dangers to humans and animals. Organic Sn, Pb, and Hg compounds are the OMC ecotoxicants that are most prevalent in the environment. Human emissions alone cannot explain the predominance of OHgC and OLCs, in fresh and marine waters, sediments, and biota, despite the fact that some of these chemicals are released directly into the environment. Lead and tin are two examples of inorganic pollutants that are discharged into the environment as OMCs with differing degrees of stability and have an impact on biota through bioconversion (Nsengimana 2007). Because of the matrix effect and their low abundance, OMCs are challenging to detect in biological and environmental samples. To improve the result, separation and preconcentration are required. Speciation studies are even more difficult, a key element that offers information on the metal's bioavailability and, in turn, its toxicity is speciation. A derivatization has been developed and successfully applied. The expected pathways for OLC and OTC have been identified after they were investigated in a range of environmental samples. A simple technique that has been devised has been used to extract them from an aqueous environmental sample. Using the novel derivatization agent from magnesium tetraethyl borate (Et_4BMgBr), an analytical technique was created for the simultaneous in situ ethylation of OTC and OHgCs in sediment samples. GC/ICP/MS and isotopic dilution and derivatization are the methods used to determine the compounds of mercury and tin. Comparing the experimental results with certified reference material (plant and sediment; $IAEA_{405}$ for sediment and CRM_{279} for plant) verified the method's accuracy and percentage recovery. While the majority of countries have banned the use of OLCs as petrol additives, some still add lead to petrol in different amounts of different TALs, and there are still locations where OLC contamination is present, such as lead alkyl manufacturers. TALs and BTCs are prohibited from use in petrol additives antifouling pints and PVC materials, respectively. This study concentrated on how they were converted in soil and water. This should shed some insight on their existence and the ways in which they are causing environmental degradation. A technique for complete speciation and identification of lead and its different forms has been devised following the breakdown of $(CH_3)_4Pb$ in aqueous samples, utilizing sodium tetraphenyl lead borate $NaB(Ph)_4$ for in situ derivatization. Direct derivatization took place in the aqueous sample, and supported liquid membrane probe extraction (SLMPE) was utilized for extracting the derivatives. The change of tetra methyl lead in aqueous solution at different concentrations of cations and anions was the main aim of such investigation. Additionally, the adsorption and desorption of ionic OLC and OTC in soil were examined. Inorganic Hg may have been methylated to MeHg in Klipriver sediments, according to the study's findings on the patterns of inorganic and methylmercury in sediment. The identification of OMCs in the environment is revitalized by recent advancements in methods and evolving speciation processes using

HPLC-MIP-AES (Chau and Wong 1991). There have been reports of further novel methods to enhance the previously challenging HPLC to furnace AAS interaction. Numerous biomedical and environmental investigations have employed GC-AAS and GC/ICP/MS in combination (Adams and Sofie 2000). Even on a regular basis, these methods reduced the detection limits to the sub-picogram level (Lobinski 1997). Among them, the extensive study by Crompton (1998) is concerned with analysis from aqueous media and almost solely concentrates on analyzing metal cations, giving limited consideration to OMCs. Similarly, metal cations are the primary focus of Ure and Davidson (1995). Despite focusing more on analytical chemistry, the work edited by Ebdon et al. (2001) takes into account speciation or the chemical forms for all the metallic compounds present, rather than only OMCs. In analytical systems that are referred to be interfaced or hyphenated, separation and detection go hand in hand. This is especially important since, despite their actual toxicity and/or environmental consequences, these species are frequently found in the environment at extremely low quantities (ppb, ppm). However, a general remark about analysis must be made. There are two things to think about: (1) the metal, OMC fragments, or entire compound must be detected by a method that is sensitive enough; (2) different forms of OMCs are present in the same samples, as each has distinct characteristics being detected separately. AAS, voltammetry, MS, AES, ICP, AFS, and X-ray and neutron activation methods are among these techniques. The primary techniques for separation are (1) capillary or traditional gas chromatography; (2) boiling point-dependent thermal desorption techniques; (3) HPLC; and (4) flow injection techniques. Nowadays, it is common practice to combine these separation and detection methods, which offers a comprehensive set of methods for analytical work that are well explained (Ebdon et al. 2001).

2.2.1 Speciation of OHgCs

There are multiple analytical processes in the methods used to determine MeHg. GC (with packed columns) and ECD, as well as acid leaching and solvent extraction, were among the earliest methods created (Quevauviller et al. 2000). Accurate analytical techniques are required to identify Hg in various matrices for risk assessment reasons. However, in work using packed columns (Hight and Corcoran 1987), it was necessary to conditioning the column with several injections of CH_3Cl to maintain its performance. In up- to- date certification program (Quevauviller 1998), capillary columns were suggested as a solution to the issue of packed columns' reduced repeatability. For the sake of method diversity (which is essential in reference material certification), the promotion of methods based on HPLC is necessary, to complement the well-established GC methods. One of the foundations of sound analytical science is the validation of reference materials. Numerous CRMs are available to confirm analytical performance and ensure that the separation procedure is accurate and devoid of systematic bias. Except for one that employed cation exchange on a polymeric column, all HPLC methods documented in the literature were based on reversed-phase techniques (Sarzanini et al. 1994). Some techniques use reversed-phase chromatography with a mobile phase made up of a pH buffer, an organic modifier, and an ion pair or chelating reagent (Table 2.1). (Chris 2000) reported that it is insufficiently sensitive to identify the concentrations of some mercury-containing species present in uncontaminated samples; thus, spike recovery tests were conducted to verify the technique. ICP/AES, ICP/MS, DCP/AES, and CV/AAS are the three photometric and plasma detection technologies that are commonly employed with HPLC. ICP/MS with sample input via a DIN was the technique with the lowest detection limits (Shum et al. 1992). ICP-MS is the only MS technique providing detection limits of 1–10 ngL^{-1}, which can be directly interfaced with conventional HPLC systems, without splitting the eluent flow to the detector. Recently, an HPLC system coupled to atmospheric pressure chemical ionization MS was used to identify methyl mercury spiked into a fish tissue CRM (DORM-1, NRCC) (Harrington et al. 1998). Because the chemical forms of different samples present are identified depended on their structural formula rather than their retention times, this system has a major advantage over elemental detection techniques. Mercury in a gas phase can be speciated using a documented technique by Schickling and Broekaert (1995).

Phenylmercury acetate (PhHgAc), diphenylmercury (Ph_2Hg), methylmercury chloride (MeHgCl), and mercury(II) chloride ($HgCl_2$) are separated by reversed-phase HPLC employing gradient elution. The ligands and the matrix were oxidized with $K_2Cr_2O_7$ before the determination. Using CVAAS, which reduces OHgC to

Table 2.1 HPLC methods for determining inorganic and organo mercury speciation in water, sediment, marine animals, and fish (Chris 2000)

Species, sample matrix, and treatment	Column name, dimensions, and manufacturer	Mobile phase	Method of detection	Sample volume (μL)	Detection limits (ng as Hg)	Reference
I. Water samples and standards						
Hg(II), MMM and MEM; open ocean CRM	Spherisorb ODS-2; 150 × 4.6 mm i.d. × 5 μm	l-cysteine (0.5%), pH 5.0; 1.6 ml min^{-1}	CV-ICP-MS (Elan 5000, PE-Sciex, Canada)	100	Hg 0.01, MMM 0.003, MEM 0.004	Wan et al. (1997)
II. Sediment, marine animals, and fish						
Hg(II) and MMM; Dorm-1 CRM (NRCC) spiked	Kromasil C18; 250 × 4.6 mmi.d. × 5 μm; HiChrom 5	Methanol (50%), water (50%), 2-mercaptoethanol (0.01%) 1 ml min^{-1}	CV-AAS (Finnigan 7000SSQ)	100 PEEK loop	Hg(II) 0.01, MMM 0.015	Harrington et al. (1998)

MMM monomethyl mercury, *MEM* monoethyl mercury

metallic Hg through a $NaBH_4$ treatment, Hg can be found. A modified simplex algorithm is used to optimize the reagent concentrations in a continuous-flow system. Mercury has detection limits set at 10.0 ng mL^{-1}. Different reactions between MeHgCl and Ph_2Hg, as well as between $HgCl_2$ and Ph_2Hg, have been shown by analysis of multi-compound mixtures. The developed method did not use either solvent extraction or chemical derivatization steps. Mercury(II) compounds were detected at a 100 ng ml^{-1} level in the gas phase. The use of cold vapor ICP-MS lowers the detection limit by about a factor of 15 in liquid samples (Wan et al. 1997), which traditional nebulization cannot accomplish. As shown in Table 2.1, the application of cold vapor AAS enables the recording of different chemical forms of down to sub nanogram levels.

2.2.2 *Speciation of OPbCs in Environmental Samples*

Rapsomanikis (1994) was the first to report the use of $NaBEt_4$ for the detection of dimethyl- and trimethyl-lead ions. The combination of larger analyte quantities (50 mL) and a cryogenic trap to capture all of the generated methyl ethyl-lead compounds allowed for the first improvement in lead detection limits. This gave an absolute detection limit of around 10 pg from 50 mL of standard solutions with a limit of quantification of around 30 pg. Data from a selection of the literature regarding the identification of alkylated species in various biological and environmental samples are provided (Van Cleuvenbergen and Adams 1992). GC/AAS was utilized to get the majority of alkylated results. The most widely used instrumental method for identifying and measuring particular alkylated chemicals is this one. Ionic alkylated and Pb^{2+} species must be derivatized to their tetra allyl (or alkyl/phenyl) analogs because GC can only separate volatile chemicals (Forsyth and Hayward 1995). Van Cleuvenbergen et al. (1990) stated that it is desirable to derivatize the ionic form of Pb, as it is has chromatographically poor separation. Either packed or capillary columns were used for the separation of OPbC species by GC, with the former being the most popular so far, possibly due to convenience in repacking, ease of handling, and large sample capacity. However, capillary columns offer far higher resolution and have been utilized in conjunction with ECD (Forsyth and Hayward 1995), where the elution of potentially interfering molecules with the alkylated species of interest is less likely due to improved separation efficiency. However, the selectivity required for the identification of lead species is typically lacking in the ECD and other common GC detectors. As a result, GC systems have been connected to a range of mass and atomic spectrometric detectors. Table 2.2 represents the most popular and sensitive gas chromatographic separation detectors (Ramaswamy et al. 2000).

MIP-OES is inherently a multielemental detection system, useful for determining essentially all elements in the periodic table using helium as a plasma gas (Zander and Hieftje 1981). This allows, in principle, the empirical elemental ratios to be determined for the species eluting from the GC system, permitting greater

Table 2.2 GC-MS, GC-ICP-MS, and GC-FID optimizing conditions used for the determination of OMCs in different environmental and biological matrices

Species/matrix	Instrumentation (LOD, ng L^{-1})	References
OLC, W	MIP-OES (0.03–0.10)	Lobinski et al. (1994), Teissedre et al. (1994), Witte et al. (1994)
OLC, W	ETAAS (8–16)	Pyrzyfiska (1996)
OLC, W	GC/ICP/MS (0.10)	Moens et al. (1997)
OLC, W	GC/MS (2.5 ng L^{-1})	Zufiaurre et al. (1997)
OLC, U	GC/MS (7.0)	Dunemann et al. (1999)
OLC, W	GC/MS (1.0–4.0)	Baena et al. (2000)
MeHg, EtHg, W	GC-MS as Hg (0.03)	Cai et al. (2000)
OLC, W	GC/MS (83–130)	Yu and Pawliszyn (2000)
MBT, DBT, TBT, MPhT, DPhT, TPh, B	GC/ICP/MS (0.05)	Rodriguez Pereiro and Carro Diaz (2002)
TBT, DBT, MBT, TPT, W	GC/FID	Cukrowska et al. (2004)
MBT, DBT, TBT, W	GC/MS	Centineo et al. (2006)

B biota, *ETAAS* electrothermal atomic absorption spectrometry, *LOD* limit of detection, *MIP-OES* microwave-induced plasma-optical emission spectrometry, *LLE* liquid-liquid extraction, *U* urine, *W* water

certainty in compound identification (Duglas and Wolfgang 1995). MIP-OES has excellent absolute detection limits for Pb, generally less than 1 pg. Using this system, Lobinski et al. (1994) achieved the highest absolute detection limits yet recorded for OPbCs, ranging from 0.02 to 0.1 pg (as Pb). Although AAS is a highly selective and sensitive method that is comparatively free from spectral interference effects, it does not have emission spectrometry's built-in multielement capabilities. It should be understood that several elements, including carrier gas flow, optical setup, atomizer length and diameter, and injection volume, are critical to getting the best performance out of the GC-AAS combo (Duglas and Wolfgang 1995). Depending on the kind of atomizer and column type, GC-AAS systems' detection limits for OPbCs can vary significantly. As the work done by Van Cleuvenbergen et al. (1990), to combine GC and AAS in a method that produces the best detection limits, the following suggestions can be made: (1) compared to packed columns, open tubular (capillary) columns offer sharper, more concentrated solute bands and higher resolution. Consequently, capillary columns provide higher peak heights for each injected analyte mass. (2) Because quartz tube atomizers normally use longer cell lengths than graphite tube systems, they typically produce higher sensitivity, which frequently increases the signal-to-noise ratio for a given analyte mass (Baxter and Frech 1993).

GC-MS is especially helpful in verifying the structure of OLCs, but it is rarely applied in Pb speciation studies. This is likely due to several factors, including its relatively high cost and lack of superior detection limits when compared to the best AAS-based detectors (Nygren and Nilsson 1987). ICP-MS has gained popularity for detection in GC, and appropriate interface designs have been described. GC-MS is a useful reference tool during method development because it makes it easier to monitor and correct for possible decomposition of Pb species during either the sample collection step or the subsequent analytical procedure.

2.2.3 *Derivatization Procedures for OTCs in Environmental Analytes*

Anthropogenic sources and minerals containing metals provide the environment's natural Sn supply (Cima 2011). Determining the Sn speciations is crucial to evaluating environmental contamination since the toxicity and biological activities of Sn are correlated with the degree of oxidation of the inorganic forms and the degree of methylation of the organic molecules (Oliveira and Santelli 2010; Cima 2011). To accurately evaluate the levels of water contamination, it is vital to ascertain the speciation of the analyte due to the varying hazards of different forms of Sn on living beings (Weber 1985; Oliveira and Santelli 2010; GDWQ 2022). Of course, majority separation techniques rely on hyphenated procedures that use gas or liquid chromatography. Because of their separation and ease of linking, MS, FPD, AAS, and AES coupled with GC techniques have been employed significantly. The literature has numerous articles examining analytical methods for determining OTCs (Donard and Pinel 1989; Morabito 1995; Quevauviller 1998). A derivatization process is necessary before the application of a GC-based approach because of the low volatility of OTCs in the environment. Derivatization also makes it possible for potential interferences to occur less frequently during the detection stage and other subsequent analytical procedures. The most popular derivatization procedures are ethylation with $NaBEt_4$, alkylation with GRs, and over-the-counter analysis with $NaBH_4$ (De La Calle-Guntinas et al. 1998). It is important to emphasize that extraction and derivatization are two of the most important stages in OTC analysis. Degradation phenomena and low derivatization yields, particularly for PhTCs, might significantly impact the quality of the results. It is required to validate, or at the very least carefully examine, the laboratory process employed for the matrix and its specific interferences (Ritsema et al. 1995). However, the absence of commercially available derivatized standards nearly invariably prevents this validation. The University of Amsterdam has recently synthesized derivatized standards for ethylating and GRs derivatization (Ariese et al. 1999). They permitted the creation of optimization and validation studies using the programs of European community-funded projects (standards, measurements, and testing program).

2.2.3.1 Hydride Generation

Volatile OTC hydrides are formed by the reaction of OTCs with an aqueous solution of $NaBH_4$, according to the following equation:

$$R_nSn^{(4-n)+} + NaBH_4 + H^+ \rightarrow R_nSnH_{(4-n)} + H_2$$

R the organic substituent and $n = 1$–3.

When AAS makes the final decision, hybridization, which can be done online or offline, is typically employed. The element under consideration and the matrix's characteristics must guide the selection of the hydride production conditions, reductant solution concentration, pH, and acid types (Donard and Pinel 1989; Abalos et al. 1997). When dealing with aqueous medium, the hydride generation approach is simple to use, permits high preconcentration factors, and gives high hybridization yields in addition to separating the analytes from any interferences (Martin and Donard 1995). When hybridization is used on increasingly complex matrices, such as sediments, biota, or even complex aquatic matrices (like wastewater), severe interferences may be seen. The hydride formation can be inhibited by high metal concentrations, which are readily found in port sediments and industrial wastes (Nakahara 1983). The presence of metals during hydride production makes BTCs susceptible to interferences, according to studies on the multielemental interference effect (Martin and Donard 1995; Ritsema et al. 1995). The strong reduction conditions linked to the $NaBH_4$ reaction, which produces metal borides, appear to be the cause of the interferences. By targeting the Sn-H bonds, these inorganic entities can most likely react with the OTC hydrides.

2.2.3.2 Alkylation by GRs

The transformation of OTCs (R_xSnX_{4-x}) is produced by reacting with GRs (RMgX) in an appropriate solvent to produce less volatile chemicals. The most used derivatization method for OTC determination is alkylation by the GR reaction, which is often accomplished by methylation, ethylation, propylation, butylation, pentylation, or hexylation (Quevauviller 1998). The mixed tetra-alkyl tins (R_xSnR_{4-x}) are among the highly stable derivatives that may be produced using this technique and are better suited for GC separation. It is important to emphasize that GRs are dangerous substances. They should only be handled by skilled professionals using the proper safety gear (gloves, glasses, etc.) because they react severely with acids, water, alcohols, ketones, etc. GR derivatization, as opposed to hydride production, enables the identification of several OTC species (methyl tins, BTCS, PhTCs) in different environmental matrices (water, sediment, and biota) with high derivatization yields and reproducibility. However, it is characterized by additional analytical steps, owing to the necessity of destroying the GRs (by quenching), and the back extraction of the mixed tetra-alkyl tins. The risk of contamination, degradation, and losses rises with the number of analytical processes. It is frequently difficult to apply the GR derivatization approach to sediments with high sulfur concentrations. In this instance, dialkyl mono-di- and trisulfides are formed when the elemental sulfur is compound extracted using OTCs and alkylated during the GR derivatization (Cai et al. 1994). The ultimate determination of the OTCs by GC-MS or

GC-FPD is hampered by these sulfur species. The oxidation of all sulfur compounds with DMD to the respective sulfones, followed by an alumina clean-up phase, appears to be a suitable strategy for removing the sulfur interference in the OTC assessment (Fernandez-Escobar et al. 1998), and the cleanup step's employment of an $AgNO_3$-coated silica column, which, after 2 h, permits the nearly quantitative removal of sulfur compounds (Schubert et al. 1998). However, in this instance, following treatment, PhTCs were eliminated. Lastly, one must take into account the fact that, in contrast to derivatized chemicals, non-derivatized ones are easily broken down, especially when they are in solution at room temperature or above. Therefore, in order to reduce the risk of decomposition, the shortest possible period between extraction and derivatization must be guaranteed (Morabito and Massanisso 2000). Several certification exercises have successfully used derivatization with various GRs on both biological and soil samples (Quevauviller 1998).

2.2.3.3 Ethylation by Sodium Tetraethyl Borate

The use of $NaBEt_4$ as a derivatizing reagent for OTCs seems to be convenient, especially in the case of aqueous samples for which direct in situ derivatization is possible, with a consequent reduction in the analytical steps (Abalos et al. 1997). The best pH for carrying out the reaction was found to be in the range 4–5. At lower pH (around 2), the formation of OTC hydrides, owing to the partial conversion of $NaBEt_4$ into $NaBH_4$, was observed (Martin and Donard 1995). A large amount of $NaBEt_4$ reagent must be used for the direct ethylating of OTCs in sediment and biological samples to achieve high yields. This is necessary to compensate for the consumption of reagents by side reactions with metals and other components in the matrices (Ceulemans and Adams 1995). $NaBEt_4$ has been used to ethylate a variety of OTCs in a straightforward one-step process (Ashby and Craig 1988). GC-AAS has successfully analyzed the volatile, completely alkylated derivatives, and GC/MS was used for identifying the resulting ethylated compounds. De La Calle-Guntinas et al. (1998) made a comparison between GR derivatization (pentylation and ethylation) and ethylation with $NaBEt_4$ on highly homogenized mussel samples. The results showed that GR prenylation and ethylation provided the highest recoveries. The main benefits of hydride production and ethylation using $NaBEt_4$ are that they are directly applicable to the samples, minimizing the many successive steps and, consequently, the potential sources of errors. These processes are especially well suited for aqueous samples. Compared to hydride generation, ethylation offers high derivatization yields for both BTCs and PhTCs. Sediment and biological samples frequently include significant interference that prevents hydride production, making it unable to take advantage of $NaBEt_4$'s primary benefit—its stability in water. Moreover, $NaBEt_4$ is unstable and breaks down when exposed to strong acids, which are frequently used to extract OTCs from solid materials. However, GRs have the benefit of potentially being used to identify different species in a different of interferences.

At present, the analytical methodologies for the determination of alkyl Pb species in essentially all sample matrices of environmental and biological interest are available. What remains to be done here is to perform interlaboratory comparisons to verify the accuracy of the diverse sample preparation and analytical production of CRMs and thus enable all workers in the field to perform quality control checks of their procedures. Some progress in this area is now being made under the auspices of the measurements and testing pogram. A clear trend in recent work on lead speciation has been the coupling of ICP-MS with various chromatographic techniques. The instrumental combination of HPLC-ICP-MS appears to be particularly attractive to GC-AAS techniques and will become of increasing importance in the future. As a variety of suitable detection systems are now available, more attention must be diverted toward improving and verifying sample preparation methodologies. Without a doubt, speciation analysis presents analysts with a significant challenge today. By following the proper protocol for the sequential extraction and use of appropriate analytical techniques and equipment, a broader use of speciation analysis in the lab can be encouraged. Since many elements' toxicity and biological activity depend on both their quantity and various chemical forms, it is imperative to comprehend elemental speciation in the modern world. Thus, speciation analysis can improve the informational value of the data.

2.2.4 *Multidimensional Method for Identifying Marine Biota Organoarsenic Species*

Owing to the interferences in biota materials, extracting the target component before electrospray MS analysis sometimes necessitates combining several separation procedures. This technique has been approved for use to characterize the As speciation in the kidney of a *T. derasa* clam (McSheehy et al. 2002). Size exclusion techniques are used through three-dimensional LC to separate the different arsenic compounds. Electrospray TOF-MS is used to examine the isolated species. The precision of the molecular mass measurement and molecular ion's isotopes enables molecule identification, which is especially important for compounds that include the monoisotopic element As. CID-MS can be used to identify unknown substances or confirm their identities. It was possible to identify twenty organoarsenic compounds without using real retention time criteria. The literature does not yet identify some of these compounds. Tandem mass spectrometry (MS), as illustrated in this example of standardless speciation analysis in an ambient biota, can yield information that is virtually identical to that of NMR, with a sample quantity that is approximately 100–1000 times smaller than that of NMR. Generally, the analytical method for speciating OMCs in a range of matrices, GC with ICP-MS detection, has matured. For common applications such as the environmental speciation of OHgC, OPbC, and OTC, it exhibits figures of merit that are comparable to those of GC-MIP-AED. However, it has several advantages when high sensitivity, multielemental screening, accurate isotope ratio measurements, or the examination of intricate interferences is needed. The development of efficient interfaces to capillary electrophoresis and HPLC, the decreasing detection limits of ICP-MS, and the increasing sensitivity of electrospray MS for molecule-specific detection at trace levels enable the characterization of endogenous organometallic species in biological systems. Even while it can be beneficial, the development of analytical tools for biochemical speciation may suffer from the lack of an interdisciplinary approach because it is carried out at the intersection of multiple study areas. Approaches to MS detection and multidimensional separation are necessary to handle the complexity of biological matrices.

2.2.4.1 Validation of an Analytical Technique Based on Hydride Generation AAS for the Measurement of Total Organic and Inorganic as in Fish

HG-AAS was used to quantify fish's total inorganic and organic arsenic. Under ideal experimental conditions, the amount of inorganic As in fish sauce ranged from 0.05 to 1.2 mg L^{-1} at the absorption of 193.7 nm, with a LOD of 0.015 mg L^{-1}. The detected total As concentrations varied from 0.03 to 2.5 mg L^{-1}, with a LOD of 0.01 mg L^{-1}. Inorganic As's recovery ranged from 97 to 1d, while organic As's range was 97–101%. This was applied to the analysis in the commercial commodities of Phan Thiet, Phu Quoc City, and Nha Trang, Vietnam (Le et al. 2021). Completely organic HG-AAS measurements of fish sauce samples were contrasted with the outcomes of HPLC-ICP/MS, a type of liquid chromatography that makes use of an ICP mill. At a 95% confidence level, the F-test and t-test demonstrated that the null hypothesis was acceptable for both variance and mean. According to the findings, the HG-AAS method, which employed basic equipment, measured the quantity of inorganic and total organic arsenic in fish sauce with good sensitivity, accuracy, and efficiency.

2.3 Increasing the Accuracy of OMC Speciation Analysis

A variety of separation methods, including distillation, extraction, and precipitation, can achieve species selectivity. Such procedures are time-consuming and often unreliable because of their often limited selectivity and sensitivity (Adams and Sofie 2000). OMCs are the easiest to quantify among the variety of compounds used in speciation analysis, as opposed to the intricate. They are frequently measured using sophisticated techniques that depend on MS or specialized and sensitive spectrometric methods for detection and capillary column GC for separation. GC-AAS, MIP-AES, and GC-MIP-AES have been employed in numerous biomedical and environmental investigations (Lobinski 1997). The rapid development of speciation techniques has brought analytical chemistry closer to the chemistry of living things by moving it from the elemental, atomic level to the world of molecular species. The detection limits have been reduced by orders of magnitude, and femtogram detection is already a reality. Specificity is also achieved when the molecular entity can be identified and given a name. These potent instruments will support and encourage more study, particularly in fields where chemistry and biology converge. Measurement methods are continuously getting better. One of the ongoing problems with speciation processes is sample preparation.

2.4 Stability of Chemical Species in Environmental Matrices

Because carbon dioxide, water production, and metal oxide have such large negative energies, all OMCs are thermodynamically unstable to oxidation. At room temperature or below, it is kinetically unstable to oxidation. This characteristic is linked to the metal atom's vacant low-lying orbitals or nonbonding electron pairs. The pace at which water assaults nucleophiles during hydrolysis is determined by the M-C bond. When the GRs are added to an aldehyde or ketone, the GR reaction produces a secondary or tertiary alcohol. Alternatively, primary alcohol is formed when formaldehyde and GRs are combined. The stability and reactivity of OMC complexes are influenced by their nature. The thermal stability of a compound decreases with increasing elemental weight in the major groups of the periodic table, which include groups 1, 2, and 13 to 15. $LiCH_3$ is more stable than KCH_3, for instance. In the absence of air, $Si(CH_3)_4$ remains stable at 500 °C, but $Pb(CH_3)_4$ breaks down quickly at the same temperature. The d-block components defy and reject this trend by strengthening and stabilizing MC bonds as one proceeds along a group. Because natural environmental samples are typically not evaluated right away after sampling, species stability in the samples is a crucial factor to take into account. The chemical species may also change significantly with prolonged storage. Numerous elements, including microbial activity, temperature, pH, light action, interactions with container material, and chemical reactions between species, can be blamed for species instability. A representative analytical speciation of the studied sample and a precise evaluation of the environmental issue are therefore accomplished by taking into account the impact of these traits on the stability of the species.

2.4.1 *Stability of Sn Species*

Typically, the analytical processes needed for the OTC assessment of environmental samples are laborious and entail multiple operational phases. Therefore, it is occasionally necessary to wait a long period between sampling and analysis. However, based on the utilization of CRMs, validation of the analytical approach is also required. As a result, both environmental and CRM samples require strict stability control. Several writers have noted that Sn compounds exhibit good stability in synthetic solutions (Dooley and Homer 1983). An interlaboratory investigation revealed that TPhT and BTC species remained stable for 3 months in water acidified with HCl, and stored in plastic bottles at 4 °C in the dark, and kept for at least 20 days (Bergamann et al. 1994; Quevauviller et al. 1996). Comez-Ariza et al. (2000) stated that four stabilization-storage methods were tested for wet sediment samples: (1) at 4 °C, (2) freezing, (3) lyophilization and refrigerator storage, and (4) storing at 4 °C while drying in a desiccator. The most reliable conservation techniques were freezing and lyophilization, as the BTC stayed stable for at least a year under all four storage conditions. However, after 3 months, a drop in the PhTC levels was recorded.

2.4.2 *Stability of Hg Species*

MeHg is extremely toxic and builds up throughout the food chain. It is considered a hazardous pollutant in the environment. Mercury's high toxicity, particularly that of organic forms, is brought on by its volatility, ease of absorption through biological membranes, and long-term removal from tissues. APHA-AWWA-WEF (1992) maintains that water samples should be collected and kept in glass bottles to determine Hg levels. Containers made of polyethylene were typically thought to be inappropriate for this use. However, it has lately been advised to sample and store potable water in 300 mL (polyethylene terephthalate) plastic bottles before a complete Hg assessment (Copeland et al. 1996). Hg was preserved for 1 week by adding 0.50 mL of 20% $K_2Cr_2O_7$ in a ratio of 1:1 with HNO_3. The stability of CH_2HgCl and $HgCl_3$ in aqueous solutions showed no detectable effects of temperature, 0 °C, and room temperature, in the dark after 3 months (Quevauviller et al. 1996). Homogenized fish in HCl with $CuSO_4$ and then separation using toluene, including MeHg and inorganic Hg, remained stable at 4 °C for 5 months in the dark (Quevauviller et al. 1993). In these biological matrices, darkness appears to be essential for the retention of mercury species. According to recent tests, MeHg in sediment samples can be stabilized by γ-irradiation for 12 months at 20 and 40 °C (Quevauviller et al. 2000).

2.4.3 *Stability of Pb Species*

The Measurement and Testing Program (BCR) has investigated the stability of OPbC in solution as part of interlaboratory investigations (Quevauviller et al. 2000). 500 ng L^{-1} of $(CH_3)_3Hg$ and $(C_2H_5)_3Hg$ in an aqueous solution are stable for 90 days, but UV irradiation, at 254 nm for 1 h, caused these species to rapidly decompose at 1.50 and 2.90 μg L^{-1} (as Pb), respectively (Van Cleuvenbergen and Adams 1992). Under comparable circumstances, dimethyl and diethyl lead species broke down more slowly, and trialkylated solutions also showed signs of deterioration in daylight, especially for triethyl lead. For a whole year, trimethyl lead remained stable in dark-stored solutions. Darkness is therefore recommended for the preservation of alkyl lead, particularly trialkyl lead. In conclusion, species stability in aqueous, sediment, and biota environmental samples is made feasible by the use of several preservation methods, including acidity, low temperature, drying, freezing, pasteurization, freeze-drying, and dark storage.

2.5 Recovery Rates for OMCs in Different Matrices

The extraction of OMCs from environmental samples is a very complex matter in which conflicting issues must be balanced: obtaining an adequate recovery, on one hand, and preventing the destruction of the compounds, on the other hand. The extraction should be done in such a way that the analyte is separated from the interfering matrix without loss, contamination, or change of the speciation and with a minimum of interferences. The quality control of analytical measurements implies that the extraction recoveries are verified, which can be done in several ways. Modern methods (gas or liquid chromatography followed by different kinds of detection like AAS, FPD, or MS) can then be used to analyze the analyte. Potential losses (such as degradation) or insufficient recovery of a particular component are among the risks associated with extraction. At the moment, adjusting the data with a recovery factor to better approximate a material's true value is how these losses are compensated for (Craig and Jenkins 2004; Yang et al. 2023).

2.5.1 Recovery from CRMs

Recovery testing by CRMs has limitations, such as the relatively small number of CRMs available for OMC assessments and the possibility that the interferences of the analyte may not be as close to the CRMs, making the recovery evaluation more dubious. Discrepancies between certified and detected values of the analyte can be ascribed to operator error, laboratory error, and inappropriate CRM use in addition to systematic mistakes (low extraction efficiency and/or derivatization yields). In this case, applying a correction factor to the results obtained on unknown samples may result in an inaccurate estimate of the true concentrations in the sample (Morabito 1995).

2.5.2 Isotopic Dilution Recovery

The most effective method of evaluating recovery is unquestionably to spike with an isotopic form of the analyte. In this instance, the chemical form and recoveries would be the same if the spike and the incurred molecule had reached equilibrium before the actual extraction. Mass spectrometry can be used independently to assess the recovery of the spike in isotope dilution for organometallic assays. The cost and accessibility of isotopically enriched OMCs and the equipment needed to detect them place restrictions on this process (Shum et al. 1992).

2.5.3 *Standard Addition Method*

The third strategy for the assessment recovery approach involves adding standards at various stages. Once more, the number of levels used may have an impact on the recovery evaluations that are achieved. The more levels there are, the more accurate the recovery evaluation is. However, the difficulties of precisely reproducing the spiking experiments, the expense of analysis, and the analysis time all rise with the number of levels (Yang et al. 2023).

2.5.4 *Internal Standards*

Assessment recoveries are often conducted using internal standards. The chemicals to be determined and the spiked compounds are chemically different in this situation, and they might not share the same chemical characteristics, though this is preferable. When many OMCs need to be established, internal standards are typically employed because it is impractical to confirm the recovery for each component separately. Naturally, there is a chance that the internal standard will behave differently from the incurred substances (both in terms of weaker binding states and different chemical properties). To sum up, differences in the nature and application of recovery investigations are significant contributors to disparities that impede the ability to compare results from speciation research. Spiking experiments are used to reduce mistakes. Even though they have drawbacks, spike tests are especially crucial (Craig and Jenkins 2004; Yang et al. 2023).

References

Abalos M, Bayona JM, Compano R, Granados M, Leal C, Prat MD (1997) Analytical procedures for the determination of organotin compounds in sediment and biota: a critical review. J Chromatogr A 788:1

Adams F, Sofie S (2000) Improving the reliability of speciation analysis of organometallic compounds. Trends Anal Chem 19(2–3):80–85

Anastas PT, Warner JC (1998) Green chemistry: theory and practice. Oxford University Press, New York

APHA-AWWA-WEF (1992) Standard methods for the examination of water and wastewater. APHA-AWWA-WEF, Washington

Ariese F, Cofino W, Gomez-Ariza JL, Kramer G, Quevauviller P (1999) Evaluation of the state-of-the-art of butyl- and phenyltin compound determinations in freshwater sediment prior to certification of a reference material. J Environ Monit 1:191–196

Armenta S, Garrigues S, de la Guardia M (2008) Green analytical chemistry. TrAC Trends Anal Chem 27:497–511

Ashby JR, Craig PJ (1988) Environmental methylation of tin: an assessment. Sci Total Environ 73(1-2):127–133

Baena JR, Cárdenas S, Gallego M, Valcárcel M (2000) Speciation of inorganic lead and ionic alkyl lead compounds by GC/MS in prescreened rainwaters. Anal Chem 72:1510–1517

Baxter DC, Frech W (1993) Coupled gas chromatography–atomic absorption spectrometry using semi-enclosed tubular atomizers: theoretical analysis of detector performance characteristics. Analyst 118(5):495

Bergamann K, Rohr U, Neidhart B (1994) Fresenius J. Anal Chim 349:815

Braun RD (2024) Chemical analysis. Encyclopedia Britannica. https://www.britannica.com/science/chemical-analysis

Cai Y, Alzaga R, Bayona JM (1994) In situ derivatization and supercritical fluid extraction for the simultaneous determination of butyltin and phenyltin compounds in sediment. Anal Chem 66:1161–1167

Cai Y, Monsalud S, Jaffe R, Jones RD (2000) Gas chromatographic determination of organomercury following aqueous derivatization with sodium tetraethylborate and sodium tetraphenylborate-comparative study of gas chromatography coupled with atomic fluorescence spectrometry, atomic emission spectrometry and mass spectrometry. J Chromatogr A 876:147–155

Castro IB, Iannacone J, Santos S, Fillmann G (2018) TBT is still a matter of concern in Peru. Chemosphere 205:253–259. https://doi.org/10.1016/j.chemosphere.2018.04.097

Centineo G, Rodriguez-Gonzalez P, Gonzalez BE, Garcia Alonso JI, SanzMede A, Cardona NF, Aranda Mares JL, Nebot SB (2006) Isotope dilution GC-MS routine method for the determination of butyltin compounds in water. Anal Bioanal Chem 384:908–914

Ceulemans M, Adams FC (1995) Evaluation of sample preparation methods for organotin speciation analysis in sediments—focus on monobutyltin extraction. Anal Chim Acta 317(1–3):161–170

Chau YK, Wong PTS (1991) Recent developments in speciation and determination of organometallic compounds in environmental samples. Fresenius J Anal Chem 339:640–645

Chen C, Chen L, Huang Q, Yu G, Lu Z, Gabrielsen GW (2024) Determination of OTCs in marine sediments from Arctic Svalbard and West Antarctic Fildes peninsula. Mar Pollut Bull 198:115845

Chris FH (2000) The speciation of mercury and organomercury compounds by using high-performance liquid chromatography. Trends Anal Chem 19(2–3):167–179

Cima F (2011) Encyclopedia of environmental health.2nd edn. Elsevier, Amsterdam, pp 351–359

Coelho-Junior GS, Borges DLG, Svoboda M, Dědina J, Kratzer J (2023) Plasma-mediated vapor generation of mercury species in a dielectric barrier discharge: direct analysis in a single drop by atomic absorption spectrometry. Spectrochim Acta B At Spectrosc 200:106596. https://doi.org/10.1016/j.sab.2022.106596

Comez-Ariza JL, Morales E, Sanchez-Rodas D, Giraldez I (2000) Statbility of chemical species in environmental matrices. Trends Anal Chem 19(2–3):200–209

Copeland DD, Facer M, Newton R, Walker PJ (1996) Analyst 121:173

Craig PJ, Jenkins RO (2004) Organometallic compounds in the environment: an overview. In: Hirner AV, Emons H (eds) Organic metal and metalloid species in the environment. Springer, Berlin. https://doi.org/10.1007/978-3-662-09135-7_1

Craig PJ, Eng G, Jenkins RO (2003) Occurrence and pathways of organometallic compounds in the environment—general considerations. In: Craig PJ (ed) Organometallic compounds occurrence in the environment2nd edn. Wiley, Oxford, pp 1–56

Crompton TR (1998) Occurrence and analysis of organometallic compounds in the environment. Wiley, Chichester

Cukrowska E, Chimuka L, Nsengimana H, Kwaramba V (2004) Application of supported liquid membrane probe for extraction and preconcentration of organotin compounds from environmental water samples. Anal Chim Acta 523:141–147

Dakova I, Yordanova T, Karadjova I (2024) Polymeric materials in speciation analysis based on solid-phase extraction. Molecules 29:187. https://doi.org/10.3390/molecules29010187

De La Calle-Guntinas MB, Scerbo R, Chiavarini S, Quevauviller P, Morabito R (1998) Comparison of derivatization methods for the determination of butyl- and phenyl-tin compounds in mussels by gas chromatography. Appl Organomet Chem 11:693

Donard OFX, Pinel R (1989) Environmental analysis using chromatography interfaced with atomic spectroscopy. Ellis Harwood, Chichester

Dooley CA, Homer V (1983) Naval ocean system. Techn. Report No. 197, San Diego, CA, p 19

Duglas CB, Wolfgang F (1995) Speciation of lead in environmental and biological samples. Pure Appl Chem 67(4):615–648

Dunemann L, Hajimiragha H, Begerow J (1999) Simultaneous determination of hg(II) and alkylated hg, Pb, and Sn species in human body fluids using SPMEGC/MS-MS. Fresenius J Anal Chem 363:466–468

Ebdon L, Pitts L, Cornellis R, Crews H, Donard OFX, Quevauviller P (2001) Trace element speciation for environment, food and health. Royal Society of Chemistry, London

Fernandez-Escobar I, Gibert M, Messeguer A, Bayona JM (1998) Complete elimination of interferences in the organotin determination by oxidation with dimethyldioxirane combined with alumina cleanup. Anal Chem 70:3703

Forsyth DS, Hayward S (1995) Further development and optimization of a quartz tube atomizer for a gas chromatography—atomic absorption spectrometer system. Fresenius J Anal Chem 351:403–409. https://doi.org/10.1007/BF00322909

GDWQ (2022) Incorporating the first and second addendaю WHO library cataloguing-in-publication.4th edn. WHO, Geneva

Gómez-Ariza JL, Morales E, Sánchez-Rodas D, Giráldez I (2000) Stability of chemical species in environmental matrices. TrAC Trends Anal Chem 19(2–3):200–209. https://doi.org/10.1016/S0165-9936(99)00192-2

Gworek B, Dmuchowski W, Baczewska-Dąbrowska AH (2020) Mercury in the terrestrial environment: a review. Environ Sci Eur 32:128. https://doi.org/10.1186/s12302-020-00401-x

Harrington CF, Romeril J, Catterick T (1998) The speciation of mercury and organomercury compounds by high-performance liquid-chromatography atmospheric-pressure ionization mass-spectrometry. Rapid Commun Mass Spectrom 12(14):911–916

Hight SC, Corcoran MT (1987) Rapid determination of methyl mercury in fish and shellfish: method development. J Assoc Off Anal Chem 70(1):24–30

Hirner AV, Emons H (2004) Organic metal and metalloid species in the environment. Springer, Berlin

Kucharski D, Stasiuk R, Drzewicz P, Skowronek A, Strzelecka A, Mianowicz K, Giebułtowicz J (2022) Fit-for-purpose assessment of QuEChERS LC-MS/MS methods for environmental monitoring of organotin compounds in bottom sediments of the Odra River estuary. Molecules 27:4847. https://doi.org/10.3390/molecules27154847

Le DV, Phan TL, Tran QH (2021) Validation of an analytical method for the determination of inorganic, organic, and total arsenic in fish sauce based on hydride generation atomic absorption spectrometry. Int Food Res J 28(6):1164–1170

Llaver M, Fiorentini EF, Oviedo MN, Quintas PY, Wuilloud RG (2021) Elemental speciation analysis in environmental studies: latest trends and ecological impact. Int J Environ Res Public Health 18(22):12135. https://doi.org/10.3390/ijerph182212135

Lobinski R (1997) Elemental speciation and coupled techniques. Appl Spectrosc 51:260A

Lobinski R, Dirks WR, Szpunar-Lobinska J, Adams FC (1994) Anal Chim Acta 286:381

Malik AK, Kaur V, Kumar S (2012) Chapter 22: metal speciation. In: Chemical analysis of food: techniques and applications. Academic, New York, pp 715–755. https://doi.org/10.1016/B978-0-12-384862-8.00022-4

Martin FM, Donard OFX (1995) Aqueous ethylation of organotin compounds in simple solution for speciation analysis by cryo-focussing and detection by atomic absorption spectrometry comparison with hydride generation. Fresenius J Anal Chem 351:230

McSheehy S, Szpunar J, Lobinski R, Haldys V, Tortajada J, Edmonds JS (2002) Characterization of arsenic species in the kidney of the clam *Tridacna derasa* by multidimensional liquid chromatography-ICPMS and electrospray time-of-flight tandem mass spectrometry. Anal Chem 74:2370–2378

Metelkova L, Zhakovskaya Z, Kukhareva G, Voskoboinikov G, Zimina O (2022) Organotin compounds (OTs) in surface sediments, bivalves and algae from the Russian coast of the Barents Sea (Kola peninsula) and the Fram Strait (Svalbard archipelago). Environ Sci Pollut Res 29:34659–34669. https://doi.org/10.1007/s11356-021-18091-0

Moens L, De Smaele T, Dams R, Van Den Broeck P, Sandra P (1997) Sensitive, simultaneous determination of organomercury, -lead, and -tin compounds with headspace solid-phase microextraction capillary gas chromatography combined with inductively coupled plasma mass spectrometry. Anal Chem 69:1604–1611

Morabito R (1995) Extraction techniques in speciation analysis of environmental samples. Fresenius J Anal Chem 351:378–385

Morabito R, Massanisso P (2000) Derivatization methods for the determination of organotin compounds in environmental samples. Trends Anal Chem 19(2-3):113–119

Morabito R, Massanisso P, Quevauviller P (2002) Derivatization methods for the determination of organotin compounds in environmental samples author links open overlay panel. TrAC Trends Anal Chem 19(2-3):113–119

Moreda-Piñeiro A, Moreda-Piñeiro J, Bermejo-Barrera P (2013) Sample pre-treatment methods for organometallic species determination.1st edn. CRC Press, Boca Raton, p 183

Nakahara T (1983) The determination of trace amounts of tin by inductively coupled argon plasma atomic emission spectrometry with volatile hydride method. Appl Spectrosc 37:539

Nsengimana H (2007) Speciation of organometallic of tin, lead, and mercury in environmental samples. Ph.D thesis, Faculty of Sciences, University of the Witwatersrand, Johannesburg, p 293

Nygren O, Nilsson CA (1987) Determination of trimethyllead salts in blood using high-resolution gas chromatography-graphite furnace atomic absorption spectrometry. J Anal At Spectrom 2:805–808

Oliveira RC, Santelli RE (2010) Occurrence and chemical speciation analysis of organotin compounds in the environment: a review. Talanta 82:9–24

Patle A, Kurrey R, Deb MK, Patle TK, Sinha D, Shrivas K (2022) Analytical approaches on some selected toxic heavy metals in the environment and their socio-environmental impacts: a meticulous review. J Indian Chem Soc 99(9):100545. https://doi.org/10.1016/j.jics.2022.100545

Pyrzyfiska K (1996) Organolead speciation in environmental samples: a review. Mikrochim Acta 122:279–293

Qi X, Li Z, Zhao C, Zhang Q, Zhou Y (2024) Environmental impacts of Arctic shipping activities: a review. Ocean Coast Manag 247:106936

Quevauviller P (1998) Method performance studies for speciation analysis. Royal Society of Chemistry, Cambridge

Quevauviller P, Darbaek I, Muntau H, Griepink B (1993) Appl Organomet Chem 7:413

Quevauviller P, Astruc M, Ebdon L, Muntau H, Cofino W, Morabito R, Griepink B (1996) Methods for environmental trace analysis. Mikrochim Acta 123:163

Quevauviller P, Marco F, Milena H (2000) Method performance evaluation for methymercury determination in fish and sediment. Trends Anal Chem 19(2-3):157–166

Ramaswamy BR, Tao H, Nakazatoa T, Miyazaki A (2000) A quantitative extraction method for the determination of trace amounts of both butyl- and phenyltin compounds in sediments by gas chromatography-inductively coupled plasma mass spectrometry. Analyst 125(10):1757–1763. https://doi.org/10.1039/B005534O

Rapsomanikis S (1994) Derivatization by ethylation with sodium tetraethylborate for the speciation of metals and organometallics in environmental samples. A review. Analyst 7:1429

Ritsema R, Martin FM, Quevauviller P (1995) Quality assurance for environmental analysis. Elsevier, Amsterdam
Rodriguez Pereiro I, Carro Diaz A (2002) Speciation of mercury, tin, and lead compounds by gas chromatography with microwave-induced plasma and atomic emission detection (GC–MIP–AED). Anal Bioanal Chem 372:74–90
Sarzanini C, Sacchero G, Aceto M, Abollino O, Mentasti E (1994) Ion chromatographic separation and on-line cold vapour atomic absorption spectrometric determination of methylmercury, ethylmercury and inorganic mercury. Anal Chim Acta 284(3):661–667
Schickling C, Broekaert JAC (1995) Determination of mercury species in gas condensates by online coupled high-performance liquid chromatography and cold-vapor atomic absorption spectrometry. Appl Organomet Chem 9:29–36
Schubert P, Fernandez-Escobar I, Rosenberg E, Bayona JM (1998) Evaluation of desulfurization procedures for the elimination of sulfur interferences in the organotin analysis of sediments. J Chromatogr 810(1–2):245–251
Shum SCK, Pang H, Houk RS (1992) On-line coupling between chromatography and inductively coupled plasma mass spectrometry current assessment. Anal Chem 64:2444
Teissedre PL, Lobinski R, Cabanis MT, Szpunar-Lobinski J, Cabanis JC, Adams FC (1994) Sci Total Environ 153:247
Ure AM, Davidson CM (1995) Chemical speciation in the environment. Blackie Academic and Professional, Glasgow
Van Cleuvenbergen RJA, Adams FC (1992) Seasonal variation of the ionic alkyllead species composition of rainwater. Environ Sci Technol 26(7):1354–1360
Van Cleuvenbergen RJA, Marshall WD, Adams FC (1990) Metal speciation in the environment. In: Broekaert JAC, Gucer S, Adams F (eds) NATO ASI series, vol 23. Springer, Berlin
Wan CC, Chen CS, Jiang SJ (1997) Determination of mercury compounds in water samples by liquid chromatography–inductively coupled plasma mass spectrometry with anin situ nebulizer/vapor generator. J Anal At Spectrom 12:683
Weber G (1985) The importance of tin in the environment and its determination at trace levels. Fresenius Z Anal Chem 321:217–224
Witte C, Szpunar-Lobinska J, Lobinski R, Adams FC (1994) Determination of trimethyl-lead in rainwater and road dust by capillary GC MIP-AE spectrometry after in situ ethylation and extraction. Appl Organomet Chem 8(7–8):621–627
Yahya M, Yaman M, Seferoğlu Z (2022) Optical chemosensors: principles, chemistry, strategies, and applications. IntechOpen, London. https://doi.org/10.5772/intechopen.105968
Yang Q, Zhao Y, Yu H, Xiong X, Huang K (2023) Recent advances of sample separation and preconcentration techniques for analytical atomic spectrometry. Appl Spectrosc Rev. https://doi.org/10.1080/05704928.2023.2231537
Yu X, Pawliszyn J (2000) Speciation of alkyllead and inorganic lead by derivatization with deuterium-labeled sodium tetraethylborate and SPME-GC/MS. Anal Chem 72:1788–1792
Zander AT, Hieftje GM (1981) Microwave-supported discharges. Appl Spectrosc 35(4):357–371. https://doi.org/10.1366/0003702814732445
Zhakovskaya Z, Neevin IA, Metelkova L, Ryabchuk DV, Kukhareva G, Grigoriev AG, Egorova A, Krek AV (2022) Mobility of metal-organic pollutants in the emerging coastal-marine sediment of the Baltic Sea: the case-example of organotin compounds in sediments of the Gulf of Finland. J Sea Res 190:102307
Zou Z, Deng Y, Hu J, Jiang X, Hou X (2018) Recent trends in atomic fluorescence spectrometry towards miniaturized instrumentation-A review. Anal Chim Acta 2019:25–37
Zufiaurre R, Pons B, Nerín C (1997) Speciation of trimethyl-lead in rainwater by gas chromatography-mass spectrometry after ethylation with sodium tetraethyl borate - comparison with other alkylation methods. J Chromatogr A 779:299–306

Chapter 3
Fate and Behavior of OMCs in the Marine Environment

Abstract OMCs are typically found in highly polar environments, including natural waters or extremely complex biological matrices, at low quantities (ng-µg). OTCs are not only found in harbors; they can also be found offshore, where contamination is less severe. This chapter follows the transformation tracks of OMCs in various habitats as well as the interactions and fate of various compartments, including a variety of internal and external circumstances. Their evolutionary modification, which fundamentally transforms their molecular identity, considerably affects their importance, usually concerning their transmission and toxicological repercussions. This chapter displays the degradation index, partitioning, as well as diffusion. Using partitioning coefficients, the biota's capacity to concentrate TBT from water, sediment, and SPM is evaluated. The biotransformation methodology is assessed as a methodical approach to be used to get beyond the practical and technical restrictions on the utilization of various OMC remediation strategies.

Keywords OMC transformation · Degradation · Partitioning · Biotransformation

3.1 Preface

Many public health and environmental protection agencies now require in-depth research on the characteristics, transformation pathways, and interactions between different environmental compartments (under a range of internal and external conditions), as well as their end-of-life and disposal, because OMC species of trace metals are frequently more problematic than their parent compounds. OMCs are present in the environmental compartments that surround us due to both human activity and their inherent occurrence as environmental products (Haydee and Dalma 2017). Numerous studies have connected the primary trends in the transportation and dispersion phenomena of OMC contaminants to their physical and chemical properties. Both inherent and external factors have been implicated in their chemical behavior, distribution, routes, and fate. While the intrinsic characteristics

T. O. Said, G. M. El Zokm, *Organometallic Compounds in the Marine Environment*, https://doi.org/10.1007/978-3-031-99189-9_3

are only connected to these parameters in addition to the partition of the pollutant, environmental factors—temperature, pH, turbulence, and advection—vary over a comparatively wide range as a function of time and place. The leaching rate of pollutants is greatly influenced by their unique geochemical properties, which include a low S value and a propensity to move away from water. BCF, which is the ratio of an organometal's level in a test biota to that of the same chemical in water, can be used to express the bio-affinity of OMCs. Some OMCs are frequently utilized in organic syntheses, but others can be found in agriculture and numerous other industries as antifouling agents, fire retardants, pesticides, and wood preservatives (Pellerito et al. 2002; Dafforn et al. 2011; Tamburri et al. 2021; Ferreira et al. 2023; Harino et al. 2023; Scianni et al. 2023; Wang et al. 2024). Chemical pollution from both organic and inorganic pollutants is a concern associated with human existence and all associated activities, including those linked to science, logistics, and tourism. Under these conditions, data on the presence and levels of the main OMC contaminants are required to evaluate the environmental condition of this sensitive and environmentally important region (Vodopivez et al. 2021). OMCs are found in a broad chemical species: nonvolatile ionic forms (like Et_3Pb^+, Bu_3Sn^+) and volatile molecular forms (like Et_4Pb, Bu_4Sn). Furthermore, some of the most challenging issues in analytical chemistry today include OMC isolation, speciation, and quantification (UNEP 1989). It has been demonstrated that a variety of living organisms suffer negative effects from nanogram levels of OTCs in different compartments of the ocean body (Meador et al. 1997; Hoch et al. 2002; Antizar-Ladislao 2008; Arai and Harino 2009; Mattos et al. 2017; Ohji et al. 2019; Bandara et al. 2021; Choi et al. 2023). Numerous environmental protection and public health organizations now demand in-depth studies of the characteristics and transformation processes in different environmental compartments (Champ 2000; Wee and Aris 2023). The fate and disposal of OMC species of trace metals, as these species are often more worrisome than their parent compounds, are not as important as the interactions and fate of these species within different environmental compartments [under different internal and external conditions]. However, the health effects seen in laboratory studies probably differ from those in natural environments because of the interplay of several environmental factors and the great diversity of organisms in natural environments (Xiaona et al. 2022). Millions of distinct OMC derivatives can therefore be produced by accounting for all known organic compounds and all electronegative metals other than the C atom. While some of them may be useful as antifouling agents, insecticides, fire retardants, wood preservatives, and in several other industries, including agriculture, many of these are widely used in organic syntheses (Pellerito et al. 2002; Batista-Andrade et al. 2018). Only a small subset of OMCs—those whose existence and toxicity have been demonstrated in marine organisms—will be the subject of this review. TBT antifouling (AF) compounds have complicated environmental behaviors in marine habitats. Since they may be ionized, the pH of seawater has a significant impact on their speciation. However, TBT is anticipated to mostly exist as TBT-OH and behave similarly to hydrophobic organic pollutants in seawater, which is generally mildly basic (Brändli et al. 2009). Clay and other suspended particles with net negatively charged surfaces will attract and absorb

TBT and other OT cations (Arnold et al. 1997; Weidenhaupt et al. 1997; Meador 2000; Hoch 2001). The acid-dissociation constant (pK_a) of TBT-OH to the TBT^+ cation in saltwater is 6.3 (Arnold et al. 1997). These fate mechanisms cause TBT to remain in fine-grained, high-organic carbon content (TOC) sediments when oxygen levels are low or depleted (i.e., hypoxic and anoxic sediment settings). TBT normally has a half-life of 1–2 years in well-oxygenated surficial sediments, but it can persist for several decades in fine-grained, anoxic, O_2-depleted sediments (Dowson et al. 1993, 1996; De Mora et al. 1995; Berge et al. 2006; Langston et al. 2015). Research indicates a positive correlation between the sediment/porewater partition coefficient (Kd) for TBT and the sediments' TOC concentration (Langston and Pope 1995). Abraham et al. (2017) reported that sediment TOC measurements, which are normally 5% TOC, are frequently used to standardize the TBT content.

3.2 Impacts, Adaptation, and Vulnerability

OMCs are employed in many different industries as process components, additives, and catalysts. Si and related elastomers are widely used in industrial applications, while OLiCs are commonly used in fine chemical synthesis. The polymerization process commonly makes use of metal complexes. Particularly, polypropylene and polyethylene are manufactured in large numbers. Zeigler's organometallic catalysts are used to create these compounds (Kumar et al. 2023).

Rapid industrialization has led to an increase in anthropogenic activities, which has elevated OMC contamination of terrestrial and aquatic ecosystems. The health of people and the environment is negatively impacted by these pollutants. Most of these contaminants are extremely poisonous and neurotoxic and cause cancer even at very low concentrations. OMC contamination has become a major worldwide issue for which there is no practical, environmentally acceptable treatment option offered by conventional methods. To reduce the amount of freely available heavy metals in the environment, scientists now believe that using microorganisms and plants is the most acceptable approach. In 1986, the USEPA estimated that 41×10^4 tons of lead were deposited into the atmosphere globally each year; however, the estimates take into account factors such as surface area, texture, and wind speed (UNEP 2003). The majority of countries have outlawed the addition of OLCs to petrol; lead is nevertheless added to petrol in certain areas in different ratios of various TAL compounds, resulting in contamination from OLCs. The manufacturing of alkyl lead and the mixing processes at refineries, garage forecourts, and gas stations are notable examples (Simon 2002). Figure 3.1 displays the map of the global distribution of leaded gasoline (Nsengimana 2007).

Significant man-made mercury emissions are found in South Africa, mostly from the burning of coal for heat and electricity (coal accounts for 90% of SA's electricity). As illustrated in Fig. 3.2 (Pacyna et al. 2006), at the moment, China and Southeast Asia are the biggest emitters. The Eastern United States, Central and Eastern Europe, and South Africa come next (Streets et al. 2009).

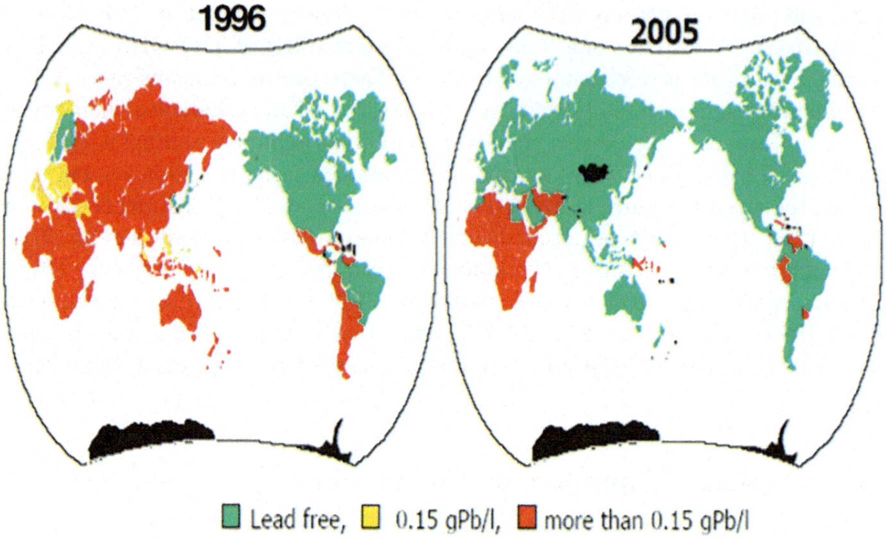

Fig. 3.1 The 1996 versus 2005 availability of leaded gasoline (Nsengimana 2007)

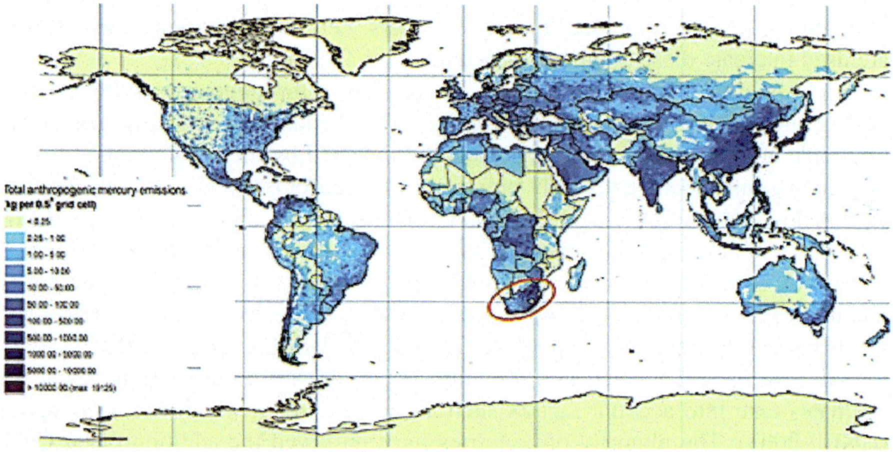

Fig. 3.2 Global anthropogenic emissions of mercury to the atmosphere, arranged geographically, 2000 (Pacyna et al. 2006)

Numerous facets of life and human welfare on Earth depend on marine and coastal environments. By controlling the cycles of heat, water, and components like carbon, the ocean, which makes up two-thirds of the planet, is also home to a wide variety of life. Marine systems provide a significant portion of the food, minerals, energy, and jobs for many nations. Our understanding of how climate change impacts ocean and coastal systems and how human societies can interact with these consequences has advanced since the last assessment due to further laboratory research, field observations and process studies, larger model simulations, and indigenous and local knowledge. An increasing amount of information demonstrates how this is felt and offers potential remedies for both ecological and human adaptation (Cooley et al. 2022). Water quality, marine life, seafood consumption, and human services are all negatively impacted by the entry of direct and indirect contaminants into the maritime environment (Datta 2023). Numerous factors can exert pressure on ecosystems, including changes in land use, the deposition of pollutants and nutrients, harvesting, livestock grazing, the introduction of exotic species, and fluctuations in the natural climate. An additional factor that could alter or jeopardize these systems is climate change. The process of development is linked to possible risks to the environment, in addition to this extensive list of difficulties. One of the main worldwide environmental concerns that has been linked to both research and politics is climate change. Degradation, loss of biological diversity, persistent organic pollutants, and stratospheric ozone depletion are further global environmental concerns (McCarthy et al. 2001). Nature contains a wide variety of OMCs and complexes. Their significance is significantly impacted by their evolutionary mutation, which completely alters their molecular identity; this is sometimes referred to as their transfer and potentially harmful adverse effects (Gao et al. 2023). The actual cytotoxic activity, the potential for traveling between living things, consistency in the visible region, high temperatures, water, and disintegration inside organisms are all governed by speciation (Pellerito et al. 2002; Warford et al. 2022). OMC use has led to serious accidents, such as those that occurred at Minamata Bay, a coastal town in Japan, as well as direct and indirect environmental input where a portion of the populace made their living from fishing. They found that the only industry that uses mercury as a catalyst is that of vinyl chloride and acetaldehyde. The Minamata Bay, which provides food for that area, is where the factory dumped its trash. Only fishermen and their families were afflicted. Over 700 of the survivors were left with serious, lasting disabilities, and 43 of the 2000 instances that were identified have since died. A thorough investigation of the Minamata catastrophe revealed that it was a textbook case of methylmercury poisoning. To assess the elements' level of toxicity, the different OMC forms of the elements in ultramicro concentrations must be continuously determined (Jantzen and Prange 1995). Human adaptation encompasses a variety of measures (adaptation choices) that mitigate the effects of climate change or take advantage of its opportunities (IPCC 2014). The people and living organisms that depend on the marine services are the main focus of adaptation strategies to mitigate major marine and coastal hazards. Reducing emissions is also necessary to mitigate the worldwide phenomenon of growing weather changes on aquatic systems; local adaptation efforts alone won't be sufficient (Oppenheimer et al. 2019). They concluded that policies are not sufficient to provide effective adaptive responses in marine ecosystems.

3.3 Environmental Significance

The existence of humans and all related activities, including scientific, logistical, and tourism-related ones, poses a risk of chemical pollution from organic and inorganic pollutants. Under these conditions, data on the presence and levels of the main OMC contaminants are required to evaluate the environmental condition of this sensitive and environmentally important region (Vodopivez et al. 2021). The fate and disposal of OMC species of trace metals, as these species are often more worrisome than their parent compounds, are not as important as the interactions and fate of these species within different environmental compartments (under different internal and external conditions). Therefore, by considering all atoms less negative than carbon and all known organic compounds, millions of different OMC derivatives can be generated. Alkyl metal derivatives are more toxic than aryl metals, while organic metal derivatives are often more dangerous than their parent inorganic metal. Furthermore, the toxicity mechanism is determined by the coordination atoms in the biological molecule that is bound. Only a small subset of OMCs—those whose existence and toxicity have been demonstrated in marine organisms—will be the subject of this review. OTCs have become a significant environmental concern in recent years, especially because of their widespread use as active ingredients in pesticide formulations (mostly TPhT) and antifouling paints (primarily TBT). As seen in the Sado Estuary in Portugal, the Crouch Estuary in the United Kingdom, and the Arcachon Bay in France, their severe toxicity to nontarget creatures like mussels and oysters, along with their direct release into the marine environment, has hurt the ecology and the economy (Chiavarini et al. 1996).

3.4 Fate of OMCs in the Marine Environment

OTCs are found in three main compartments of aquatic ecosystems: the water column, the upper layer of sediments, and the surface microlayer. Tributyl tin degrades easily to di- and mono-butyl. A concentration of 0.03 μg^{-1} of TBT declined from a clean water site had half-lives between 9 and 19 days for light and dark treatments, respectively (Dubey and Roy 2002). The rate and pattern of TBT's sequential degradation into dibutyl-tin, mono-butyl tin, and Sn (IV) were described using a first-order multistep kinetic model (Sarradin et al. 1995). Abiotic degradation processes are additional potential pathways for TBT breakdown from soil, sediments, and water columns. The Sn-C bonds may be attacked by such abiotic processes in some ways. Examples include chemical split, γ-irradiation, thermal split, and breakdown aided by UV radiation. Only when exposed to UV light (300–350 nm), with around 300 kJ mol^{-1}, is TBT expected to undergo direct photolysis. Since UV light does not penetrate deeply, photolysis should only occur in the first few centimeters of the water column. Maureen and Willingham (1996) asserted that the TBT degradation process is defined as the sequential disappearance of an alkyl group from TBT to create hazardous inorganic Sn, as can be seen directly below:

$$R_3Sn^{+} \rightarrow R_2Sn^{2+} \rightarrow RSn^{3+} \rightarrow Sn(IV) \tag{3.1}$$

Despite its low solubility and restricted mobility, TPT has a propensity to adsorb to soil and silt in the aquatic habitat (Blunden et al. 1986). Alkali metals, carboxylic acids, and mineral acids can all mediate chemical cleavage for unbound organotin that is accessible by chemical action. These substances can cleave Sn-C bonds heterolytically using both electrophilic and nucleophilic processes (Blunden and Evans 1990).

3.4.1 The Origin of Contaminants and the Degradation Index

One of the powerful properties of OMCs is their ability to break down into inorganic molecules, which are essentially nontoxic. Strong degradation methods include biological activities like biotransformation and physicochemical mechanisms like hydrolysis and photodegradation (Gomez et al. 2024). They stated that TBT typically breaks down via physicochemical and biological processes, including hydrolysis and photodegradation, as well as biotransformation. OMCs break down in bottom sediments over several years, which is much slower than in water (Dowson et al. 1996). By computing a degradation index, one can evaluate the quantitative significance of TBT deterioration and forecast its current contamination (Diez et al. 2002). Assuming comparable rates of degradation for both chemicals, the BDI and PhDI were computed using the following Eqs. (3.2) and (3.3):

$$BDI = \frac{[MBT] + [DBT]}{[DBT]} \tag{3.2}$$

$$PhDI = \frac{[MPhT] + [DPhT]}{[TPhT]} \tag{3.3}$$

The BDI of less than 1 is indicative of fresh TBT contamination (Fig. 3.3).

The association between TBT and inorganic Sn in sediments was further examined using a modified method (BDImod; Eq. (3.4)). This method considers the inorganic Sn as the degradation's final product (Furdek Turk et al. 2020):

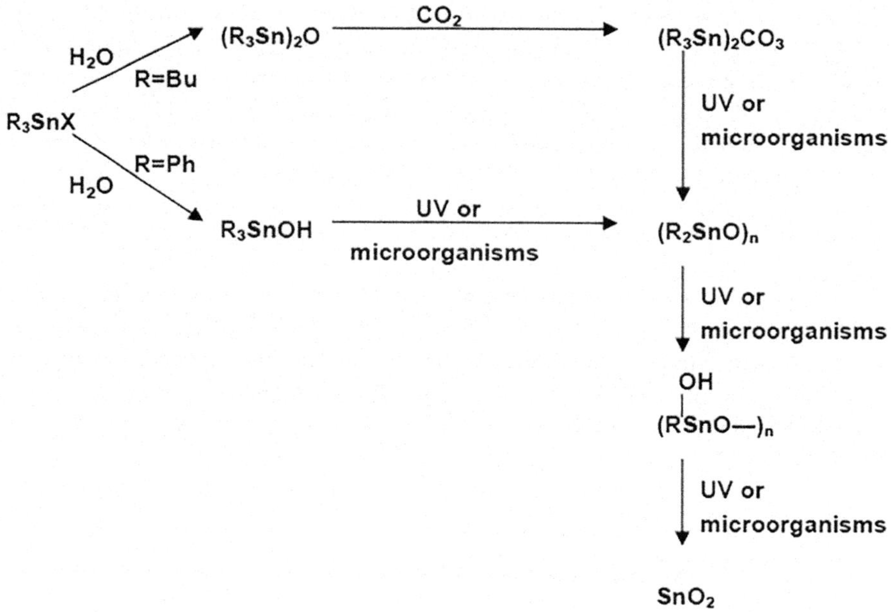

Fig. 3.3 TBT and TPT compound environmental degradation scheme (Sunday et al. 2012)

$$\mathrm{BDI}_{\mathrm{mod}} = \frac{\left(\mathrm{MBT} + \mathrm{DBT} + Sn_{\mathrm{inorganic}}\right)}{\mathrm{TBT}} \quad (3.4)$$

3.4.1.1 Partitioning of TBT in Marine Environment

One of the most significant mechanisms lowering the levels and hazardous of TBT in the ecosystems is thought to be sorption (Fent 1996). Therefore, apparent distribution coefficients for sediment-water ($k_{s/w}$), mussel-water ($k_{m/w}$), mussel-sediment ($k_{m/s}$), mussel-SPM ($k_{m/\mathrm{SPM}}$), water-SPM ($k_{m/\mathrm{SPM}}$) are recorded. The ratio of TBT concentrations in the various matrices was computed to evaluate their distribution in seawater, sediments, mussels, and SPM ($k_{s/\mathrm{MPS}}$). It is possible to assess mussels' capacity to concentrate TBT from water, sediment, and SPM using the coefficients $k_{m/w}$, $k_{m/s}$, and $k_{m/\mathrm{SPM}}$, respectively (Quintas et al. 2021).

Low concentrations of TBT and TPhT, ranging from ng L^{-1} in water to μg kg^{-1} in sediments, have been found to have detrimental impacts on a range of living creatures (Meador et al. 1997). The shell length growth of juvenile *Mytilus edulis* was measured at 24–48 h intervals for 7 days after they were exposed to 0.10–10.00 μg L^{-1} of TBTO in flowing raw seawater. The length growth rate significantly decreased at TBTO concentrations greater than 0.400 μg L^{-1} (Stromgren and Bongard 1987). Growth response and TBTO concentration have a roughly hyperbolic relationship. It has been demonstrated experimentally that Atlantic salmon

(*Salmo salar*) accumulate elevated levels of tin in various organs from exposure to low concentrations (0.10–1.00 μg L^{-1}) of TBT in solution; commercially grown salmon from farms using TBT-treated nets typically contain approximately 0.50–1.00 mg kg^{-1} TBT in mussel tissue. Antifouling paints containing TBT compounds have been used sparingly on netting cages at Scottish coastal salmon production facilities (Davies and Mckie 1987). OMCs are utilized for a wide range of applications: manufacturing of pharmaceutical drugs like cisplatin; use for agricultural applications like manufacturing of ferrocene which is utilized for soil remediation, crop protection, and herbicides and fungicides; used as stoichiometric reagents in research-oriented and industrial chemical reactions; and employed as catalysts to increase reaction rates (e.g., in the use of homogeneous catalysis). OMCs help transmit oxygen in the circulation and convert carbon dioxide to oxygen and organic matter. They are used in bulk hydrogenation operations, such as the making of margarine. OMCs are used to form various other organic compounds. Along the Catalan coast, a thorough investigation on the prevalence, biochemical impacts, and effects of OTC contamination on fish and mollusks has been conducted (Morcillo et al. 1997; Morcillo and Porte 1998). The concentration of TBT in mollusks remains high even after legislation to curb TBT inputs was passed, according to chemical examination of biota samples. Sediments and live species close to important Portuguese commercial harbors recorded high contents of TBT. Additionally, it was demonstrated that TBT levels, >25 μg L^{-1}, have an impact on the larval species of *Ruditapes decussatus* (Langston et al. 1997). Unpalatable concerns have been raised by the rising percentage of OTC in environmental samples. A range of OTC applications and sources, including environmental levels, toxicity, and analytical methods of determination, have been reviewed. There is a need to conduct a thorough environmental assessment of OTCs in both phases of marine water (water and sediment), as well as biota samples, to determine the concentration levels, due to the lack of enforcement regarding the control of them in the marine habitat, particularly in Africa (Okoro et al. 2011). OTCs are not only found in harbors; they can also be found offshore, where contamination is less severe. Several fish species from the Western Mediterranean have been found to have both TBT and TPhT (Kannan et al. 1996). OTCs have been found in marine creatures all around the world for a number of years after they were regulated, and the imposex phenomena in gastropods have been well documented. It is yet unknown what processes underlie the imposed reaction and whether there are any more sensitive biochemical precursors to this phenomenon. A more thorough assessment of OTCs' effects on coastal habitats would require more research. The use of TBT-based antifouling paints is nearly always prohibited by legislation since the leaching of TBT is now acknowledged globally hazard to the life (Bosselmann 1996). Mineral cinnabar contains the majority of Hg, with an average crustal abundance of 0.05–0.10 mg kg^{-1} on Earth (Fergusson 1990). Sea spray, geysers, wind-borne dust, volcanic emissions, and thermal fluids are examples of natural sources and transportation methods (Lobinski and Marczenko 1996). Paints, electrical items, dentistry uses, and the organochlorine pesticide industry have all made substantial use of Hg. Approximately half originate from the electrolysis of brine to produce chlorine and caustic soda (Craig 1986). Mercury is now a worldwide contaminant that has been detected in the deep ocean, Antarctica's atmosphere, and the Arctic due to its extensive use and the erratic behavior of certain species (Fergusson 1990). The tendency of Hg to be methylated in the environment, becoming MMM and DMM, is one of the most fascinating and significant aspects of its biogeochemistry. Field and laboratory research by Hamasaki et al. (1995) has demonstrated that microorganisms methylate Hg(II), although the exact role of abiotic processes is yet unknown. Temperature, pH, redox conditions, and sulfide concentration are the primary variables influencing the process. MeHg makes up around 2% of the total Hg in saltwater and 0.10–1.50% of the total Hg in sediments. Nonetheless, it is magnified in marine life, and it makes up almost all (~80.0%–90.0%) of the total Hg in fish. Because MMM and the other OHgCs, such as MEM and MPhM, are lipophilic and can penetrate biological membranes to enter cells, they are more hazardous than mercury (II) (Craig 1986). Once within the cell, mercury can attach to biomolecules with thiol groups and disrupt a number of metabolic functions. However, Pb has been utilized in a wide range of products since at least the time of the Bible, primarily in inorganic form, thus until this century, its effects as an environmental contaminant were negligible. Pb pollution has greatly increased as a result of its use in storage batteries, metallic form, and as organic anti-knocking additives (TAL) in gasoline (Duglas and Wolfgang 1995). TAL compounds, which can pass through biological membranes and the skin and be easily absorbed through the lungs, are approximately ten times more poisonous to the nervous system than inorganic Pb. In the environment, TAL pollutants are broken down into trialkyl lead ions by sunlight and atmospheric components (like ozone and the hydroxyl radical), while in the biological system, dealkylation takes place through reactions with thiol groups in proteins and enzymes. Subsequent conversion produces inorganic lead through alkylated intermediates (Radojevic and Harrison 1987).

Inorganic As has been the subject of toxicological evaluations and human exposure to date, whereas organic As has typically been regarded as nontoxic. Pellerito et al. (2002) examined how OMCs are toxic to marine life (Fig. 3.4). Methylated As compounds, which are created when enzymes methylate inorganic As (iAs) to produce molecules with one to four methyl groups, are found in marine environments. Dimethyl arsenate (DMA) is the most prevalent of these compounds, which are typically found in seafood as minor As species. Higher amounts of DMA (3–46%) than those found in finfish or algae have been reported in molluscs (Fricke et al. 2004; Cleland et al. 2009; Moreda-Pineiro et al. 2010; Whaley-Martin et al. 2012; Berges-Tiznado et al. 2013; Taylor et al. 2017).

Inorganic As despite having very low concentrations (less than 0.01 μg kg^{-1}), (arsenate and arsenide), are among the most hazardous types (Buchet et al. 1996). The three primary organic forms of arsenic are arsenobetaine, arsenosugars, and arsenolipids (Mania et al. 2015). Although their quantities are similar to those of iAs, methyl arsenate (MA) and DMA are the most dangerous organic arsenic compounds present in seafood (Lorenzana et al. 2009). While other organic arsenic species identified in seafood are known to be nontoxic, some species and their intermediate metabolites are cytotoxic in cell cultures (Meyer et al. 2014, 2015;

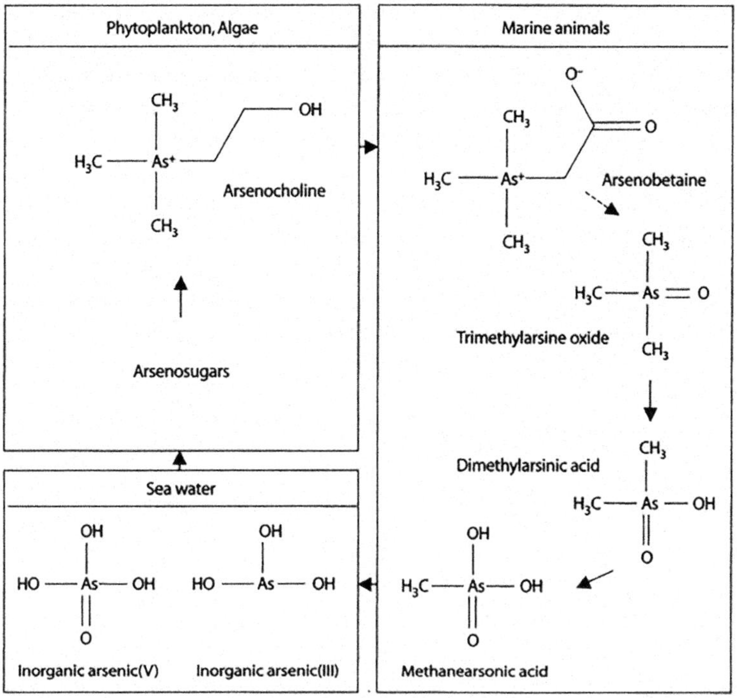

Fig. 3.4 A tentative arsenic cycle in marine ecosystems (Pellerito et al. 2002)

Molin et al. 2015; Taylor et al. 2017). In marine environments, monomethyl As (MMAs) are rare and typically only found in trace amounts; however, it is present in highly recordable levels in certain biota species (Kirby and Maher 2002; de la Calle et al. 2011). Tetramethyl arsonium (TETRA) are recorded in gastropods at high levels (Francesconi and Kuehnelt 2004) and clams (Shiomi et al. 1987). As illustrated in Fig. 3.5a, the As compounds found in seafood are present in many forms based on their chemical identity and characteristics (EFSA 2009; Julshamn et al. 2012). Figure 3.5b shows the movement of arsenic from lower to higher trophic levels; bioaccumulation and biomagnification of arsenic are indicated by the red dots (Ghosh et al. 2022). Figure 3.6a shows the estimated As species distributions, which are explained in detail below (Julshamn et al. 2012). Estimates are derived from a review of the literature and the average levels for various marine foods (EFSA 2009; Julshamn et al. 2012). It is observed that the evaluation of organic As in marine foods is limited by a few studies that demonstrate their speciation, as well as the discrepancies in analytical methodologies across those studies.

Figure 3.6b illustrates the consumption of various seafood classes from various countries with different diets (EFSA 2009). Estimates of the distribution of As species and median seafood concentrations were then used to calculate the possible consumption of As species by various consumer groups (Fig. 3.6c). Seafood's organic As species concentrations have an impact on exposure as well (EFSA 2009).

Numerous bivalves contained dissolved As compounds. A methanol/water (1:1) solution was used for their extraction and then analyzed using an HPLC/ICP/MS (Shibata and Morita 1992). *T. japonica* and *T. keenae* contain both arsenobetaine and arsenosugar derivatives. Given that these bivalves consume plankton, algae may be the source of the aforementioned arsenosugars. In contrast, arsenobetaine may be mistakenly accumulated by mussels due to its molecular resemblance to glycine betaine, which is utilized by the same organisms for osmoregulation. Given that *Corbicula japonica* is a mussel inhabiting low-salinity areas, the lack of arsenobetaine in its diet may be caused by a decrease in the mussel's osmoregulation. Microorganisms found in sediments have been shown to cause the synthesis of arsenobetaine from arsenocholine in vitro (Hanaoka et al. 1997).

a

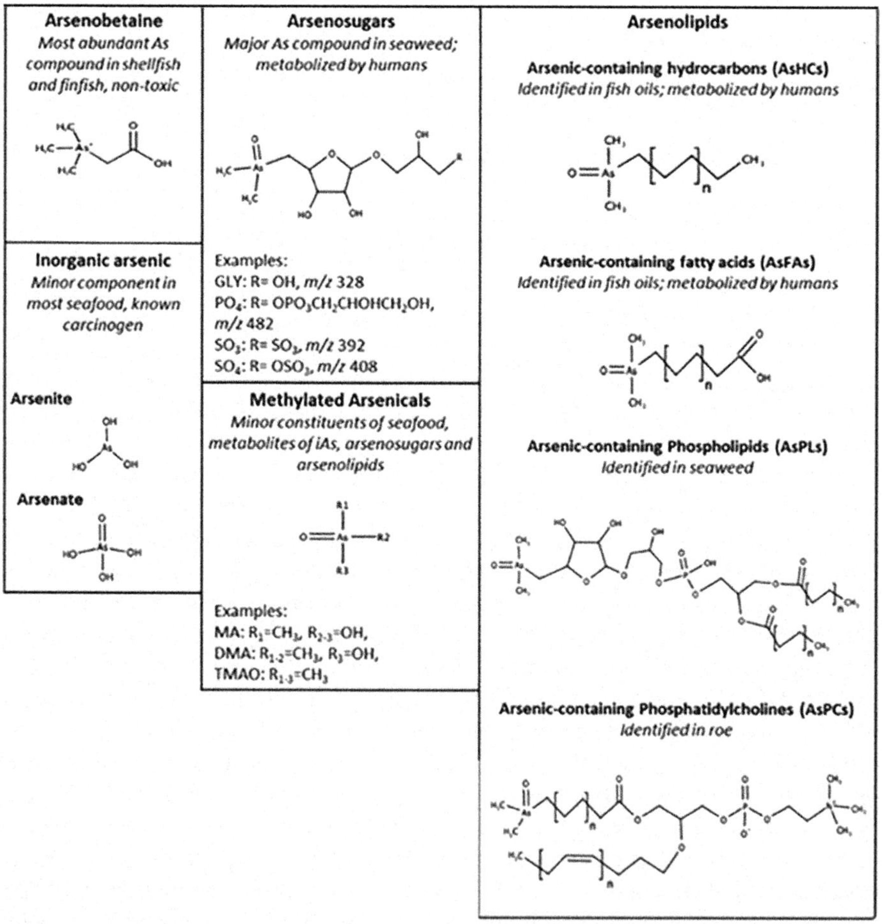

Fig. 3.5 (**a**) Arsenic compounds found in seafood (EFSA 2009; Julshamn et al. 2012). (**b**) Diagram of the movement of arsenic from lower to higher trophic levels; the red dots indicate bioaccumulation and biomagnification of arsenic (Ghosh et al. 2022)

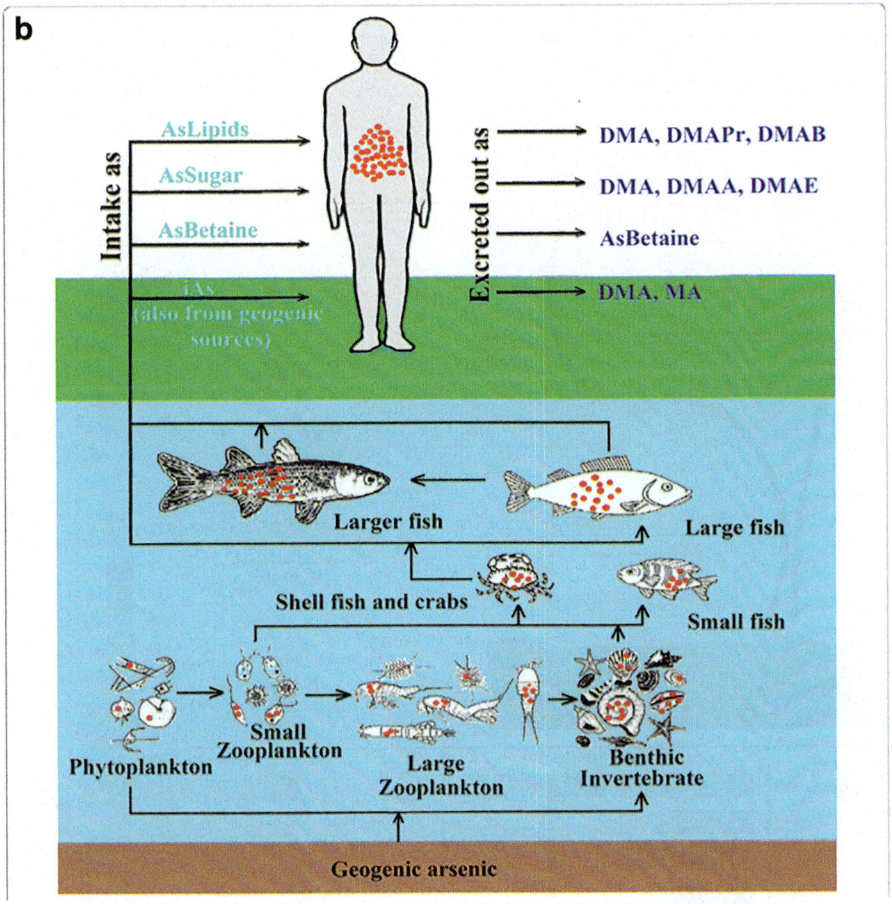

Fig. 3.5 (continued)

3.4.2 An Overview of OTC Fate

3.4.2.1 OTC Derivatives

For a variety of reasons, organotin (IV) derivatives have been the subject of much research. Examples of their widespread use around the world include PVC stabilizers, agrochemicals, organochlorinated pesticides, agricultural preservatives, and antifouling agents (algaecides, molluscicides) used in marine environments. In recent years, books and reviews have been written that discuss their identity of these substances as well as nearly all of their uses (Patai 1995; Champ and Seligman 1996; Smith 1998; Barbieri et al. 2000). Consequently, sediments, marine biota, and natural and marine waters all contain significant amounts of organotin(IV) compounds. There have been many excellent books recently published that discuss the

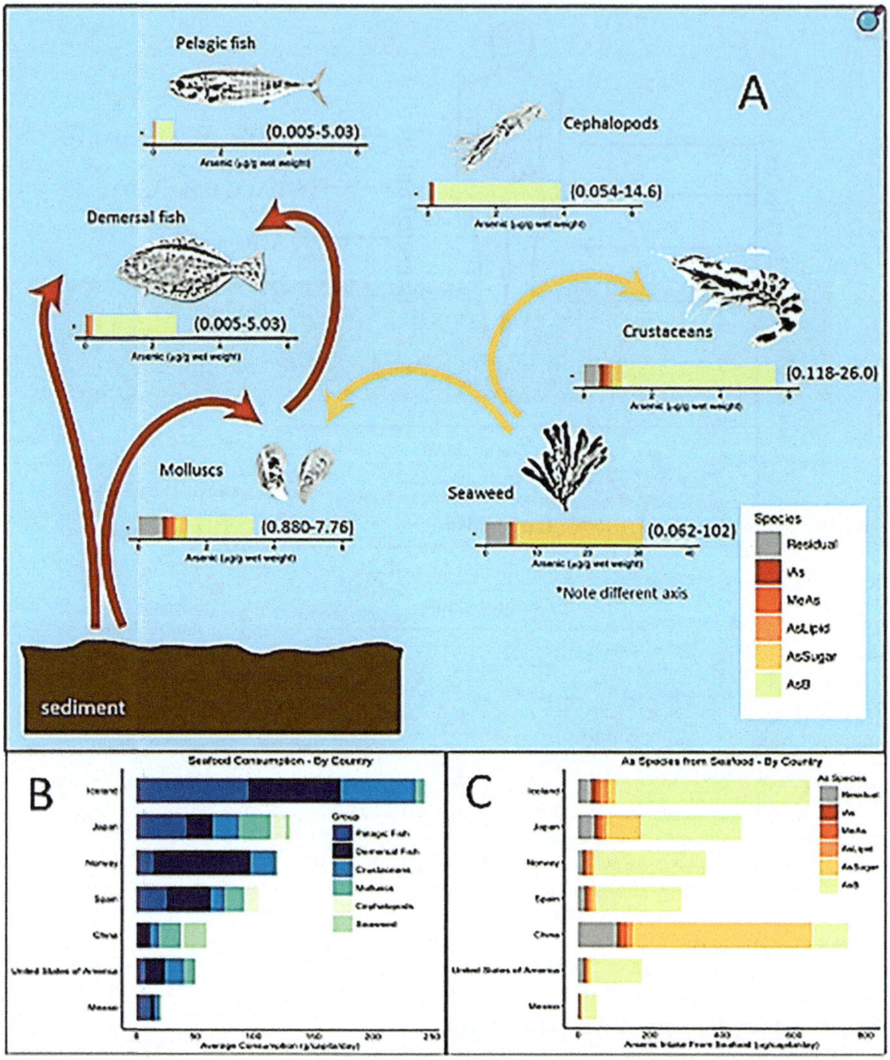

Fig. 3.6 (**a**–**c**) As species intake scenarios from various seafood diets. (**a**) Approximated As species distributions in various seafood varieties. Plotted demersal vs. pelagic fish concentrations were estimated from the EFSA (2009) mean As concentration for fish and the ratio of mean As in demersal to pelagic fish from the Norwegian fish survey (Julshamn et al. 2012). Total As concentrations are based on mean concentrations from EFSA (2009). In brackets is the range of total As concentrations (5–95% confidence limit). (**b**) Food Consumption Database-based seafood consumption by nation. (**c**) Assumptive consumption of As species derived from data on seafood consumption and estimated distributions of As compounds (EFSA 2009)

toxic effects of OMCs and include derivatives of organotin(IV) (Arakawa and Wada 1993; Mennie and Craig 1993; Crompton 1998). Toxic effects of OTC degradations on habitats have been recorded in the last decade to stay within the scope of this issue. In addition, the degradation of OTCs in the ocean is presented in Fig. 3.7 (Zhang et al. 2021).

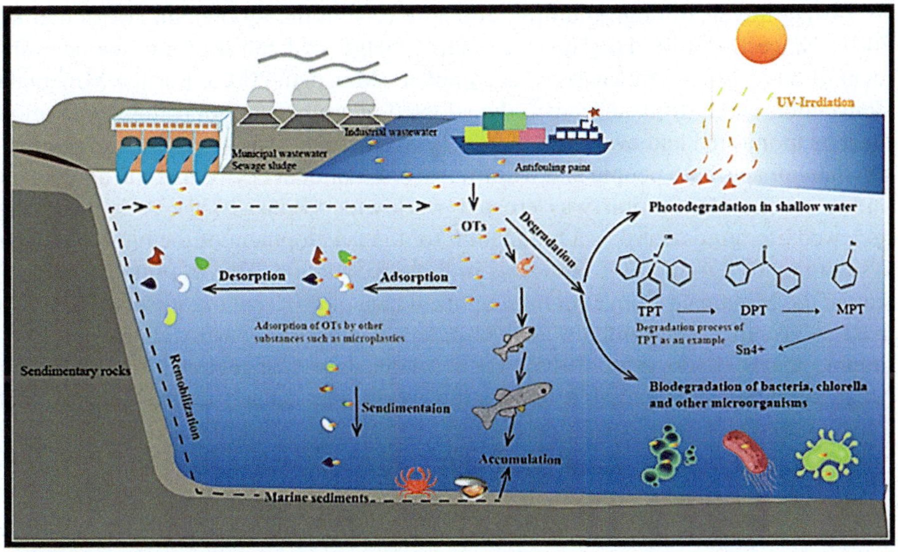

Fig. 3.7 Ocean organotin degradation (Zhang et al. 2021)

3.4.2.2 OTC in the Marine Biota

TBTO has been detected in sediments and *Mytilus galloprovincialis* samples collected in Taranto Port, which has highly established mussel cultures. Four distinct sampling stations were used in the extraction procedure, along with the n-C_6H_6/CH_2Cl_2 mixture extraction and a NaOH wash for extract purification. The data were obtained via electrothermal AAS. According to Cardellicchio et al. (1992), the amounts of TBTO found in mussels and sediments ranged from 11.0 to 30.0 and 15.0 to 47.0 ng g^{-1} wet weight, respectively.

3.4.3 An Overview of Organo Mercury Behavior

Mercury can exist in two most significant organometallic forms: DMHg [$(CH_3)_2Hg$] and MMHg [$(CH_3)Hg^+$] (Devai et al. 2005; Zhong and Wang 2008). Volatile species like elemental Hg and DMHg cause transportation from soil to atmosphere, which is a crucial route for Hg (Bohme et al. 2005). Mercury is said to be methylated by bacteria that reduce sulfate. The addition of sulfate can boost the synthesis of methylmercury (Jeremiason et al. 2006). However, the availability of Hg^{2+} for methylation is reduced when sulfide is present (Kongchum et al. 2006). Inorganic Hg precipitates as a highly insoluble HgS mineral due to the buildup of reduced sulfur, mostly as sulfide (Kaplan et al. 2002). The observed reduced methylating activity in marine and estuary sediments compared to freshwater sediments may be due to the salinity factor (Kongchum et al. 2006).

MMHg travels through multiple trophic levels up the food chain (Ullrich et al. 2001). Since DMHg[$(CH_3)_2Hg$] is a volatile OHgC that can leak into the environment, it has a lower propensity to accumulate (Manahan 2000). For the Minamata Convention on Hg to be effective, the adopted measures and their implementation must be monitored and evaluated. After Hg was identified as an environmental poison that impacts both people and wildlife, an international environmental agreement known as the Convention was created worldwide (Bank 2020). A lot of peer-reviewed literatures serve as a foundation for the development of the biomonitoring system required to monitor Hg levels in biota worldwide. They reported Hg exposure in the Minamata Convention-recognized taxa of different organisms. They also used to visualize Hg contents for continents and oceanic basins at relevant geographic scales (Evers et al. 2024). Their results show that regional biomonitoring programs were largely deficient over the world, particularly in the Atlantic and Pacific Oceans, Africa, Australia, the Indo-Pacific, and the Middle East. They did, however, discover a few useful regional simulations for controlling MeHg concentrations in their aquatic systems.

3.4.4 Organo Lead Fate

Despite a wealth of information on the detrimental effects of Pb, there is still debate on the impacts of occupational exposure, minor-level exposure, and the historical use of Pb in many purposes. With over 60 years of use as gasoline anti-knock additives, OPbCs occupy a prominent position (Patocka 2008). The majority of industries that were previously associated with elevated Pb levels were present. However, incidences of severe exposure and even direct lead poisoning still happen. The majority of focus is currently being paid to the effects of exposure that are not yet apparent, as blood Pb levels have dropped below the threshold at which symptoms or indicators arise in the majority of businesses. It is frequently discussed how significant some of these effects are for the workers' overall health. Since recent research indicates that no level of lead exposure appears to be safe and even the low levels of exposure currently in place are linked to neurodevelopmental deficits, especially in children, there is unavoidably pressure to reduce Pb exposure in the general population and workplace. Pb is highly present in blood levels. Low birth weight babies and miscarriages have been linked to it even at lower levels (Nordstrom et al. 1979). Blood Pb levels above 10.00 μg/100 ml are linked to a greater chance of low birth weight, early delivery, and spontaneous miscarriage in women exposed to Pb at work. In one investigation, Pb blood levels between 50.0 and 90.0 $\mu g\ L^{-1}$ were associated with an increased likelihood of spontaneous miscarriage (Borja-Aburto et al. 1999). According to recent research, low libido, low count of semen, and abnormal morphology may all be related to blood Pb levels in males over 40 μg/100 ml, which may affect reproductive function (IPCS (International Programme on Chemical Safety) 1995). Many studies, however, have not considered social factors like smoking, alcohol consumption, or medication use, or other occupational exposures (heat, solvents, etc.) that could be powerful confounders. A predefined technique was used to examine semen samples from lead workers in Belgium, Italy, and the United Kingdom. The results indicated that men with blood Pb levels above 500 μg L^{-1} had a 49% decrease in median sperm concentration, with a realistic threshold for impacts of 440 μg L^{-1}. Additionally, there was modest evidence of chromatin degradation in sperm among the men whose spermatozoa had the greatest quantity of Pb. Data from biological monitoring showed that Pb had no long-term effects on chromatin or sperm quality (Bonde et al. 2002). Blood Pb levels exceeding 400 μg L^{-1} may alter the shape and function of sperm, while those below about 500 μg L^{-1} are unlikely to significantly affect reproductive ability. To offer a more accurate estimate of the tolerable threshold of Pb exposure, more study on male and female workers is needed. The majority of OLCs are becoming less and less common, especially in light of the nearly complete elimination of Pb alkyls from petrol globally. Tetraethyl Pb production peaked in the 1970s at around 5.0×10^5 tons worldwide, but it is currently estimated to be around 4.5×10^3 tons worldwide. They were widely used as octane enhancers. Although they used to have the highest risk of illness and even death, the only groups that are rarely exposed are workers who clean storage tanks that have housed leaded gasoline and those who produce, transport, and blend this material into gasoline. This will be an issue for some time, though, as tanks that have previously held leaded gasoline will require cleaning or repairs, as well as continuous care. Since tetraethyl lead primarily causes hazardous toxicity, its toxicological profile differs from that of inorganic Pb and its derivatives (Gidlow 2015). Early symptoms are mild and ambiguous, making them hard to identify. However, serious symptoms of toxic psychosis or even coma and death may arise from repeated exposure or a single, large dose. Tetraethyl Pb has been categorized by the IARC as a category three carcinogen, meaning that there is not enough data to support the classification. However, one published study suggests that production workers may be more susceptible to rectal cancers than other workers (Fayerweather et al. 1997). Tetraethyl Pb is changed by the liver into soluble alkyl Pb chlorides, which the body then eliminates through urine. While the blood Pb level may only be slightly elevated in cases of organic Pb poisoning, the urine Pb level may be considerably elevated, with several hundred micrograms of Pb per gram of creatinine. As a result, in questionable situations, urinary lead levels should always be examined; otherwise, the diagnosis might not be made. There is currently no known cure for organic Pb poisoning; supportive care and appropriate sedation are the only treatments available. In contrast, chelating drugs may be utilized in the event of poisoning from inorganic Pb and its compounds.

The Health and Safety Executive (2002) was concerned about levels of Pb in workers in the United Kingdom. They outline the obligations of employers, occupational doctors, and workers. The physicians who monitor Pb workers are either HSE employees or medical professionals designated by law to act on their behalf. Workplace familiarity, annual testing of Pb workers, tracking of their blood Pb levels, and annual reporting of blood Pb levels to the HSE are among the responsibilities. Additionally, if a worker has blood Pb levels above the suspension limit or shows any indications of Pb toxicity, they must be included in the investigation. The designated physician must also advise the employer to notify the HSE of any suspected symptoms of Pb poisoning and notify the HSE of any employee who has been suspended from exposure to either organic or inorganic Pb by RIDDOR, given by the Health and Safety Executive (1995). The CLAW Regulations maintain the 0.60 mg L^{-1} suspension threshold for male Pb workers and the 300 μg L^{-1} suspension standard for female workers with reproductive capability. However, lead poisoning symptoms could appear at lower blood lead levels. A 0.30 mg L^{-1} suspension limit for men was proposed by SCOEL; however, the socioeconomic effects on the sector must be examined prior to the implementation of any such limit (European Commission 2002). As the author is aware, the European Commission designated SCOEL to assess their 2003 conclusion that the health-based blood lead limit ought to be set at 0.300 mg L^{-1} and to conduct a follow-up assessment of the health effects of Pb shortly. SCOEL may look at more current data and suggest a far lower threshold. Nonetheless, a lot of nations, including the European Union, have passed laws lowering the suspension thresholds for both male and female employees. No employee's blood Pb level shall surpass 0.300 mg L^{-1} for men and 0.010 mg L^{-1} for women who are capable of reproducing if no action is taken, according to a voluntary aim set by the European Battery Association, the International Lead Association, and the European Lead Sheet Association, REACH Regulations, a formal risk assessment of Pb toxicity was conducted (Health and Safety Executive 2008). Kosnett et al. (2007) examined a noteworthy publication that affected the conclusion of the REACH assessment; this study was a meta-analysis of research that had been done on the effects of Pb on health (Table 3.1). Some firms that expose their employees to Pb in other ways have also voluntarily accepted these suspension restrictions, as have other Pb users. An interesting paradox exists at the moment: while workers exposed to Pb who are not technically Pb users (such as demolition and cleanup workers) are nonetheless subject to the higher suspension limits contained in the CLAW standards, Pb users have accepted industry-agreed suspension limits derived under the REACH standards.

It's interesting to note that laws are beginning to consider the long-held belief in the Pb industry that Pb levels in the air are not particularly essential and that a person's blood Pb level is mostly determined by their cleanliness and the proper use of personal protective equipment. It is well known that cigarette smokers and nail biters have significantly higher blood lead levels than other workers. Because employers are required by the CLAW Regulations to pay workers while they are suspended due to high blood Pb levels, it will be necessary to increase the frequency of worker monitoring to ensure that blood Pb levels are not approaching the suspension limit. Additionally, approved analytical laboratories will need to respond with ever-increasing accuracy and precision of results (Table 3.2). Ensuring that females with reproductive potential stay below the 0.01 mgL^{-1} mg/L suspension limit will present a unique challenge for the industry (Health and Safety Executive 2002).

Table 3.1 Blood lead levels and potential effects on adult health (Kosnett et al. 2007)

Blood Pb level ($\mu g\ L^{-1}$)	Males	Females
<5.0	Nil	Nil
5.0–10.0	Hypertension, kidney dysfunction	Hypertension, kidney dysfunction, and spontaneous abortion
11–20	Hypertension, kidney dysfunction, and subclinical neurocognitive deficits	Hypertension, kidney dysfunction, subclinical neurocognitive deficits, reduced birth weight, and postnatal developmental delay
21–29	Hypertension, kidney dysfunction	Hypertension, kidney dysfunction
	Subclinical neurocognitive deficits	Subclinical neurocognitive deficits; spontaneous abortion; reduced birth weight; postnatal developmental delay
30–39	Hypertension, kidney dysfunction, and neurocognitive deficits	Hypertension, kidney dysfunction, neurocognitive deficits, spontaneous abortion, reduced birth weight, postnatal developmental delay
40–79	Hypertension; kidney dysfunction; subclinical peripheral neuropathy; Neurocognitive deficits; anemia; sperm abnormalities; colic; gout	Hypertension; kidney dysfunction; subclinical peripheral neuropathy; neurocognitive deficits; anemia; colic; gout; spontaneous abortion; reduced birth weight; postnatal developmental delay
80+	Hypertension; nephropathy; peripheral neuropathy; neurocognitive deficits; anemia; sperm abnormalities; colic; gout; encephalopathy	Hypertension; nephropathy; peripheral neuropathy; neurocognitive deficits; anemia; colic; gout; encephalopathy; spontaneous abortion; reduced birth weight; postnatal developmental delay

3.5 Long-Term Reduction of Effluents

The redox reaction changes the valence of OMCs, reducing their toxicity, in microbial and phytoremediation. Additionally, research into the use of biochar as a remediation technique has produced a viable option that merits wider application. Enzymes that are environmentally effective biocatalysts for eliminating OMCs from contaminated areas are known to be secreted by microbes and whole microbial cells. As far as we are aware, there is very little literature on the remediation process for heavy metals and the sensors utilized for heavy metal detection. To get beyond the practical and technological restrictions on using these bioremediation strategies, systematic management should be put in place (Gaur et al. 2021). It is now difficult to reduce or eliminate hazardous heavy metals. There are three categories of heavy metal treatment: chemical, biological, and physicochemical techniques (Hashim et al. 2011). *P. aeruginosa*, *P. jamilae*, *B. subtilis*, *Aspergillus* sp., *Botrytis* sp., and *Neurospora* sp. demonstrated to degrade a variety of heavy metals (Gururajan and Belur 2018). Different processes, such as phytovolatilization in *Chara canescens*, *Brassica juncea*, and *Typha latifolia*, allow plants to eliminate heavy metals (LeDuc and

Table 3.2 Current international suspension limits (Health and Safety Executive 2002)

Country	PbB µg/dl	
	Males	Females
Germany	40	10
Finland	50	50
Denmark	20	20
Sweden	50	30
Italy	60	40
France	40	30
Belgium	70	70
Spain	70	70
Portugal	70	70
Netherlands	70	70
UK	60	30
Ireland	70	70
Switzerland	70	30
Hungary	50	30
Slovenia	40	30
Poland	50	50
Australia	30	10
Japan	40	40
China	40	
India	40	
South Africa	40	

Terry 2005; Sakakibara et al. 2012). Both *Populus alba* and *Morus alba* activate phytoextraction and phytoaccumulation as reported by Awa and Hadibarata (2020). However, rhizofiltration of Phaseolus vulgaris and Helianthus annuus was reported by Cristaldi et al. (2017). The muscle tissues of European eels have been found to bioaccumulate several metals, including mercury (0.360), arsenic (0.230), copper (0.920), cadmium (0.011), and lead (0.047) in mg kg^{-1}, wet weight (Rudovica and Bartkevics 2015). Furthermore, biochar, biosurfactant, and biocatalytic removal are some of the notable advancements in the field of heavy metal removal that have recently occurred (Gaur and Manickam 2021; Wang et al. 2021). Biosurfactants are secondary metabolites of microorganisms that can contain a variety of xenobiotic pollutants, including heavy metals (Gaur et al. 2019), be able to solubilize (Tripathi et al. 2020; Mishra et al. 2021). Previous studies of cleanup attempts based on microorganisms and plants have been excellent (Varjani 2017). Nevertheless, there is a dearth of published works that summarize the most recent heavy metal remediation interventions. Therefore, in addition to phyto and micro-based heavy metal remediation techniques, we have covered in detail the latest developments in heavy metal reduction and sensor technologies. Its limitations have been analyzed in connection to the knowledge gaps and potential future directions.

3.6 Biotransformation

All microorganisms, both prokaryotic and eukaryotic, interact with various metal classes. It is also known that some microbes exhibit catalytic activity when exposed to the metalloid selenium (Craig 1986). The interactions involved are more biological than environmental, and these processes need incredibly low elemental concentrations. By redox, methylation, and demethylation, microorganisms can also change metal and metalloid species. Since altered metal chemical forms can alter an element's mobility, toxicity, and physical and chemical properties, many of these biotransformations have significant environmental effects. Several bacteria, such as *Cyanobacterium*, *Enterobacter cloacae*, and *Desulfovibrio vulgaris*, have been demonstrated to enzymatically convert Cr(VI) to Cr(III) (Garnham and Green 1995). It is believed that the reduction involves at least two steps and involves Cr(V). Waterways contaminated with Cr(VI) can be cleaned using chromium bioreduction since Cr(III) is far less dangerous (Garnham and Green 1995). Adsorption also causes Cr(III) to get immobilized in anaerobic sediments (Moore 1990), in line with bioremediation. Both SeO_4^{-2} and SeO_3^{-2} can be converted by bacteria to elemental selenium, even though only the former has been shown to encourage growth. The NO_3^- inhibits SeO_4^{-2}, not SeO_4^{-2}, according to reports, suggesting that nitrate reductase plays a role in dissimilatory selenium reduction. Nevertheless, it has been demonstrated that the bacteria *Thauera selenatis* may grow anaerobically by reducing SeO_4^{-2} through certain reductases that are different from those engaged in NO_3^- reduction. Likewise, the facultatively anaerobic bacteria *Enterobacter cloacae* simultaneously lower NO_3^- and respires SeO_4^{-2} in anaerobic circumstances (Losi and Frankenberger 1997).

During respiration, some bacteria employ U(VI) as an outer electron acceptor, enabling them to proliferate and produce U(IV). Acetate is used as an electron donor by *Geobacter metalli*, whereas *Shewanella putrefaciens* connects hydrogen oxidation to U(VI) reduction (Lovley et al. 1993). Both bacteria are known to reduce Fe, in addition to U(VI) (Lovley 1995). Nonetheless, *Desulfovibrio* species lower U(VI), but at a rate that is out of proportion to development (Lovley and Phillips 1992; Lovley et al. 1993), meaning that energy is not saved during this reductive process. *D. vulgaris* uses cytochrome c3 to reduce U(VI) (Lovley et al. 1993). Abdelouas et al. (1998) reported the reduction of uranium in underground water with high nitrate, sulfate, and carbonate levels through biocatalysis. Microbially driven reactions caused a gradual decrease, taking place simultaneously with both SO_4^- and S^- reduction. Many researchers have reported on the biotransformation of arsenic in maritime environments (Francesconi et al. 1992; Kaise and Fukui 1992; Garman et al. 1993). Phytoplankton only discriminates between arsenate and phosphate by a factor of two to ten in concentration differences because the concentrations of the two substances may be almost equimolar in the coastal boundary zone. As a result, organoarsenic chemicals such as arsenobetaine, dimethyl arsinic acid, and methane arsonic acid account for the majority of the arsenic that builds up in marine species. Figure 3.8 discusses OMCs' biotransformation, dispersion, and fate

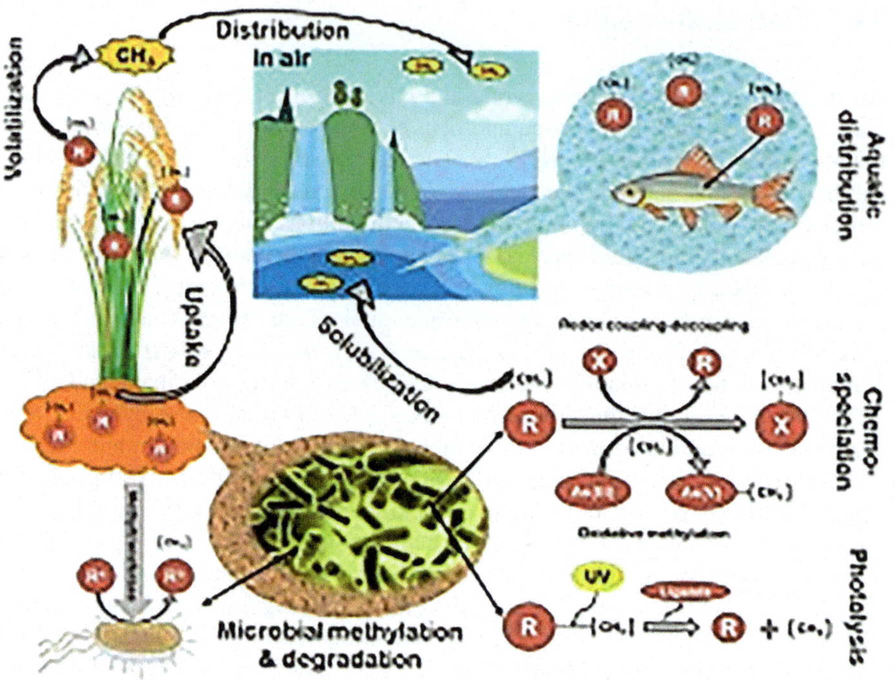

Fig. 3.8 Distribution, reactions, and biotransformation of OMCs in aquatic environments (Majumdar et al. 2024)

in aquatic environments (Majumdar et al. 2024). Many bacterial taxa, such as *Cryptococcus*, *Pseudomonas*, and *Staphylococcus* species, and many intestinal bacteria may convert inorganic Hg [Hg(II)] into metallic Hg [Hg(O)] (Ullrich et al. 2001). The capacity to decrease Hg(II) is frequently linked to mercury resistance; the reduction takes place during aerobic development and is believed to be a detoxifying process. The *cytosolic flavoprotein* mercuric reductase (MR), an enzyme linked to mercury resistance, depends on nicotinamide adenine dinucleotide phosphate (NADPH), a necessary electron donor in all living things, to supply the reducing power for anabolic reactions and redox equilibrium (Freedman et al. 2012).

References

Abdelouas A, Yongming L, Lutze W, Nuttall HE (1998) J Contam Hydrol 35:217

Abraham M, Westphal L, Hand I, Lerz A, Jeschek J, Bunke D, Leipe T, Schulz-Bull D (2017) TBT and its metabolites in sediments: survey at a German coastal site and the central Baltic Sea. Mar Pollut Bull 121:404–410

Antizar-Ladislao B (2008) Environmental levels, toxicity, and human exposure to tributyltin (TBT)-contaminated marine environment. A review. Environ Int 34:292–308

Arai T, Harino H (2009) Contamination by organotin compounds in Asia. In: Arai T, Harino H, Ohji M, Langston WJ (eds) Ecotoxicology of antifouling biocides. Springer, Tokyo. https://doi.org/10.1007/978-4-431-85709-9_4

Arakawa Y, Wada S (1993) Biological properties of alkyltin compounds. In: Sigel H, Sigel A (eds) Metal ions in biological systems, vol XXIX. Marcel Dekker Inc., New York, pp 101–136

Arnold CG, Weidenhaupt A, David MM, Muller SR, Haderlein SB, Schwarzenbach RP (1997) Aqueous speciation and 1-octanol-water partitioning of tributyl- and triphenyltin: effect of pH and ion composition. Environ Sci Technol 31:2596–2602

Awa SH, Hadibarata T (2020) Removal of heavy metals in contaminated soil by phytoremediation mechanism: a review. Water Air Soil Pollut 231:1–15

Bandara KRV, Chinthakac SDM, Yasawardened SG, Managea PM (2021) Modified, optimized method of determination of Tributyltin (TBT) contamination in coastal water, sediment, and biota in Sri Lanka. Mar Pollut Bull 166:112202

Bank MS (2020) The mercury science-policy interface: history evolution and progress of the Minamata Convention. Sci Total Environ 722:137832

Barbieri R, Pellerito L, Ruisi G, Barbieri-Paulsen A, Barone G, Posante S, Rossi M (2000) 119Sn Mossbauer spectroscopy studies on the interaction of organotin(IV) salts and complexes with biological systems and molecules. In: Gianguzza A, Pelizzetti E, Sammartano S (eds) Chemical processes in marine environments. Springer, Berlin, pp 229–244

Batista-Andrade JA, Caldas SS, Batista RM (2018) From TBT to booster biocides: levels and impacts of antifouling along coastal areas of Panama. Environ Pollut 234:243–252

Berge JA, Amundsen CE, Eggen T, Hylland K, Bøe E (2006) Naturlig nedbryting og biotilgjengelighet av tinnorganiske forbindelser i marine sedimenter. Norwegian Institute for Water Research (NIVA), Oslo, p 72

Berges-Tiznado ME, Paez-Osuna F, Notti A, Regoli F (2013) Biomonitoring of arsenic through mangrove oyster (Crassostrea corteziensis Hertlein, 1951) from coastal lagoons (SE Gulf of California): occurrence of arsenobetaine and other arseno-compounds. Environ Monit Assess 185(9):7459–7468

Blunden SJ, Evans CJ (1990) Organotin compounds. In: Hitzinger O (ed) The handbook of environmental chemistry, vol 3, Part E. Anthropogenic compounds. Springer, Berlin, pp 1–44

Blunden SJ, Hobbs LA, Smith PJ, Craig PJ (eds) (1986) Organometallic compound in environment. Longman, Harlow, p 111

Bohme F, Rinklebe J, Staerk HJ, Wennrich R, Mothes S, Neue HU (2005) A simple field method to determine mercury volatilization from soils. Environ Sci Pol 12:133–135

Bonde JP, Joffe M, Apostoli P, Dale A, Kiss P, Spano M, Caruso F, Giwercman A, Bisanti L (2002) Sperm count and chromatin structure in men exposed to inorganic lead: lowest adverse effect levels. Occup Environ Med 59:234–242

Borja-Aburto VH, Hertz-Picciotto I, Rojas Lopez M, Farias P, Rios C, Blanco J (1999) Blood lead levels measured prospectively and risk of spontaneous abortion. Am J Epidemiol 150:590–597

Bosselmann K (1996) Tributyltin: case study of environmental contaminants. Cambridge University Press, Cambridge

Brändli RC, Breedveld GD, Cornelissen G (2009) Tributyltin sorption to marine sedimentary black carbon and amended activated carbon. Environ Toxicol Chem 28:503–508

Buchet JP, Lison D, Ruggeri M, Foa V, Elia G (1996) Assessment of exposure to inorganic arsenic, a human carcinogen, due to the consumption of seafood. Arch Toxicol 70:773–778

Cardellicchio N, Geraci S, Marra C, Paterno P (1992) Determination of tributyltin oxide in coastal marine sediments and mussels by electrothermal atomic absorption spectrometry. Appl Organomet Chern 6:241–246

Champ MA (2000) A review of organotin regulatory strategies, pending actions, related costs, and benefits. Sci Total Environ 258:21–71

Champ MA, Seligman PF (1996) Organotin: environmental fate and effects. Chapman and Hall, London

Chiavarini S, Cremisini C, Morabito R (1996) Caroli S (ed) Element speciation in Bioinorganic chemistry. Wiley, New York

Choi HC, Lee JW, Hwang UK, Jeon HJ, Oh SY, Kim CW, Kang HS (2023) Effects of tributyltin-contaminated aquatic environments and remediated water on early development of sea urchin (Hemisentrotus pulcherrimus). Animals 13:3078. https://doi.org/10.3390/ani13193078
Cleland B, Tsuchiya A, Kalman DA, Dills R, Burbacher TM, White JW, Faustman EM, Marien K (2009) Arsenic exposure within the Korean community (United States) based on dietary behavior and arsenic levels in hair, urine, air, and water. Environ Health Perspect 117(4):632–638
Cooley S, Schoeman D, Bopp L, Boyd P, Donner S, Ghebrehiwet DY, Ito SI, Kiessling W, Martinetto P, Ojea E, Racault MF, Rost B, Skern-Mauritzen M (2022) Oceans and coastal ecosystems and their services. In: Climate change 2022: impacts, adaptation and vulnerability. contribution of working group II to the sixth assessment report of the Intergovernmental Panel on Climate Change. Cambridge University Press, Cambridge, pp 379–550. https://doi.org/10.1017/9781009325844.005
Craig PJ (1986) Organometallic compounds in the environment. Longman, London
Cristaldi A, Oliveri Conti G, Hea Jho E, Zuccarello P, Grasso A, Copat C (2017) Margherita ferrante, phytoremediation of contaminated soils by heavy metals and PAHs. A brief review. Environ Technol Innov 8:309–326. https://doi.org/10.1016/j.eti.2017.08.002
Crompton TR (1998) Occurrence and analysis of organometallic compounds in the environment. Wiley, Chichester
Dafforn KA, Lewis JA, Johnston EL (2011) Antifouling strategies: History and regulation, ecological impacts and mitigation. Mar Pollut Bull 62:453–465
Datta MG (2023) Pollution in marine ecosystem: impact and prevention. In: Soni R, Suyal DC, Morales-Oyervides L, Fouillaud M (eds) Current status of marine water microbiology. Springer, Singapore. https://doi.org/10.1007/978-981-99-5022-5_8
Davies IM, Mckie JC (1987) Accumulation of total tin and tributyltin in muscle tissue of farmed Atlantic salmon. Mar Poll Bull 18(7):405–407
de la Calle MB, Emteborg H, Linsinger TPJ, Montoro R, Sloth JJ, Rubio R, Baxter MJ, Feldmann J, Vermaercke P, Raber G (2011) Does the determination of inorganic arsenic in rice depend on the method? TrAC Trends Anal Chem 30(4):641–651
De Mora SJ, Stewart C, Phillips D (1995) Sources and rate of degradation of tri(n-butyl)tin in marine-sediments near Auckland, New-Zealand. Mar Pollut Bull 30:50–56
Devai I, Patrick WH, Neue HU, De Laune RD, Kongchum M, Rinklebe J (2005) Methyl mercury and heavy metal content in soils of rivers Saale and Elbe (Germany). Anal Lett 38:1037–1048
Diez S, Ábalos M, Bayona JM (2002) Organotin contamination in sediments from the Western Mediterranean enclosures following 10 years of TBT regulation. Water Res 36(4):905–918. https://doi.org/10.1016/S0043-1354(01)00305-0
Dowson PH, Bubb JM, Lester JN (1993) Depositional profiles and relationships between organotin compounds in fresh-water and estuarine sediment cores. Environ Monit Assess 28:145–160
Dowson PH, Bubb JM, Lester JN (1996) Persistence and degradation pathways of tributyltin in freshwater and estuarine sediments. Estuar Coast Shelf Sci 42:551–562
Dubey SK, Roy U (2002) Biodegradation of tributyltins (organotins) by marine bacteria. Appl Organometal Chem 17:3–8
Duglas CB, Wolfgang F (1995) Speciation of lead in environmental and biological samples. Pure Appl Chem 67(4):615–648
EFSA (2009) Scientific opinion on arsenic in food. EFSA panel on contaminants in the food chain. EFSA J 7:198
European Commission (2002) Recommendations from the Scientific Committee on occupational exposure limits for lead and its inorganic compounds. SCOEL/SUM/83
Evers DC, Ackerman JT, Åkerblom S et al (2024) Global mercury concentrations in biota: their use as a basis for a global biomonitoring framework. Ecotoxicology 33:325–396. https://doi.org/10.1007/s10646-024-02747-x
Fayerweather WE, Karns ME, Nuwayhid IA, Nelson TJ (1997) Case-control study of cancer risk in tetraethyl lead manufacturing. Am J Ind Med 31:28–35
Fent K (1996) Ecotoxicology of organotin compounds. Crit Rev Toxicol 26(1):1–117

Fergusson JE (1990) The heavy elements: chemistry, environmental impact and health effects. Pergamon Press, Oxford

Ferreira ADM, Monteiro de Matos J, Silva LK, Martins Viana JL, Freitas MDSD, Júnior OPA, Santos Franco TCR, Brito NM (2023) Assessing the spatiotemporal occurrence and ecological risk of antifouling biocides in a Brazilian estuary. Environ Sci Pollut Res. https://doi.org/10.1007/s11356-023-31286-x

Francesconi KA, Kuehnelt D (2004) Determination of arsenic species: a critical review of methods and applications, 2000–2003. Analyst 129(5):373–395

Francesconi KA, Edmonds JS, Stick RV (1992) Arsenocholine from anaerobic decomposition of a trimethylarsonioriboside. Appl Organomet Chern 6:247–249

Freedman Z, Zhu C, Barkay T (2012) Mercury resistance and mercuric reductase activities and expression among chemotrophic thermophilic Aquificae. Appl Environ Microbiol 78(18):6568–6575. https://doi.org/10.1128/AEM.01060-12

Fricke MW, Creed PA, Parks AN, Shoemaker JA, Schwegel CA, Creed JT (2004) Extraction and detection of a new arsine sulfide containing arsenosugar in molluscs by IC-ICP-MS and IC-ESI-MS/MS. J Anal At Spectrom 19(11):1454–1459

Furdek Turk M, Ivanić M, Dautović J, Bačić N, Mikac N (2020) Simultaneous analysis of butyltins and total tin in sediments as a tool for the assessment of tributyltin behaviour, long-term persistence and historical contamination in the coastal environment. Chemosphere 258:127307. https://doi.org/10.1016/j.chemosphere.2020.127307

Gao J, You J, Wu J, Guo J, Fu P, Zhang L (2023) Factors affecting the accumulation of organotins by wild fish: a case study in the Three Gorges Reservoir, China. Environ Sci Pollut Res 30:124407–124415

Garman GD, Pillai MC, Cherr GN (1993) Inhibition of cellular events during early algal gametophyte development: effects of select metals and an aqueous petroleum waste. Aquat Toxicol 28(1–2):127–144

Garnham GW, Green M (1995) Chromate (VI) uptake by and interactions with cyanobacteria. J Ind Microbiol 14:247–251. https://doi.org/10.1007/BF01569935

Gaur VK, Manickam N (2021) Microbial biosurfactants: production and applications in circular bioeconomy. In: Pandey A, Tyagi RD, Varjani S (eds) Biomass, biofuels, biochemicals. Elsevier, Amsterdam, pp 353–337

Gaur VK, Bajaj A, Regar RK et al (2019) Rhamnolipid from a Lysinibacillus sphaericus strain IITR51 and its potential application for dissolution of hydrophobic pesticides. Bioresour Technol 272:19–25

Gaur VK, Sharma P, Gaur P, Varjani S, Ngo HH, Guo W, Chaturvedi P, Singhania RP (2021) Sustainable mitigation of heavy metals from effluents: Toxicity and fate with recent technological advancements. Bioengineered 12(1):7297–7313

Ghosh D, Ghosh A, Bhadury P (2022) Arsenic through aquatic trophic levels: effects, transformations and biomagnification—a concise review. Geosci Lett 9:20

Gidlow DA (2015) Occup Med 65(5):348–356. https://doi.org/10.1093/occmed/kqv018

Gomez NA, Sturla Lompré J, Ferrando A, Garrido M, Domini CE (2024) Update on the status of the contamination by organotin compounds in sediment of Nuevo Gulf, Argentina. Insights from field and experimental studies. Mar Pollut Bull 200:116087. https://doi.org/10.1016/j.marpolbul.2024.116087

Gururajan K, Belur PD (2018) Screening and selection of indigenous metal tolerant fungal isolates for heavy metal removal. Environ Technol Innov 9:91–99

Hamasaki T, Nagase H, Yoshioka Y, Sato T (1995) Crit Rev Environ Sci Technol 25(1):45–91

Hanaoka K, Kaise T, Kai N, Kawasaki Y, Miyasita H, Kakimoto K, Tagawa S (1997) Arsenobetain-decomposing ability of marine microorganisms occurring in particles collected at depths of 1100 and 3500 meters. Appl Organometal Chern 11:265–272

Harino H, Ohji M, Kono K, Onduka T, Hano T, Mochida K (2023) Current status of antifouling biocides contamination in the Seto Inland Sea, Japan. Arch Environ Contam Toxicol 85:333–348

Hashim MA, Mukhopadhyay S, Sahu JN et al (2011) Remediation technologies for heavy metal contaminated groundwater. J Environ Manag 92:2355–2388

Haydee KM, Dalma KE (2017) Concerning organometallic compounds in environment: occurrence, fate, and impact. In: Recent progress in organometallic chemistry. IntechOpen, London

Health and Safety Executive (1995) RIDDOR regulations. HSE, London

Health and Safety Executive (2002) Control of lead at work regulations HSE. HSE, London

Health and Safety Executive (2008) Research, evaluation, authorisation and restriction of chemicals regulations

Hoch M (2001) Organotin compounds in the environment - an overview. Appl Geochem 16:719–743

Hoch M, Alonso-Azcarate J, Lischick M (2002) Adsorption behavior of toxic tributyltin to clay-rich sediments under various environmental conditions. Environ Toxicol Chem 21(7):1390–1397

IPCC (2014) Annex II: glossary. In: Climate change 2014: impacts, adaptation, and vulnerability. Part B: regional aspects. Cambridge University Press, Cambridge, pp 1757–1776

IPCS (International Programme on Chemical Safety) (1995) Inorganic lead. Environmental health criteria. World Health Organization, Lyon

Jantzen E, Prange A (1995) Organometallic species of the elements tin, mercury and lead in sediments of the longitudinal profile of the River Elbe. Fresenius J Anal Chem 353:28–33

Jeremiason JD, Engstrom DR, Swain EB, Nater EA, Johnson BM, Almendinger JE, Monson BA, Kolka RK (2006) Sulfate addition increases methylmercury production in an experimental wetland. Environ Sci Technol 40:3800–3806

Julshamn K, Nilsen BM, Frantzen S, Valdersnes S, Maage A, Nedreaas K, Sloth JJ (2012) Total and inorganic arsenic in fish samples from Norwegian waters. Food Addit Contam Part B Surveill 5(4):229–235

Kaise T, Fukui S (1992) The chemical form and acute toxicity of arsenic compounds in marine organisms. Appl Organomet Chern 6:155–160

Kannan K, Corsolini S, Focardi S, Tanabe S, Tatsukawa R (1996) Accumulation pattern of butyltin compounds in dolphin, tuna and shark collected from Italian coastal waters. Arc Environ Contam Toxicol 31:19–23

Kaplan DI, Knox AS, Myers J (2002) Mercury geochemistry in wetland and its implications for in situ remediation. J Environ Eng 128:723–732

Kirby J, Maher W (2002) Tissue accumulation and distribution of arsenic compounds in three marine fish species: relationship to trophic position. Appl Organomet Chem 16(2):108–115

Kongchum M, Devai I, De Laune RD, Jugsujinda A (2006) Total mercury and methylmercury in freshwater and salt marsh soils of the Mississippi river deltaic plain. Chemosphere 63:1300–1303

Kosnett MJ, Wedeen RP, Rothenberg SJ, Hipkins KL, Materna BL, Schwartz BS, Hu H, Woolf A (2007) Recommendations for medical management of adult lead exposure. Environ Health Perspect 115:463–471

Kumar A, Avula B, Yadav S, Singh Y, Eraganaboyina S, Zyryanov GV, Nemallapudi BR, Singh AK, Bhimireddi R (2023) Synthesis, properties, environmental stability and practical application of organometallic compounds: a comprehensive review. Inorg Chem Commun 158(2):1387–7003. https://doi.org/10.1016/j.inoche.2023.111567

Langston WJ, Pope ND (1995) Determinants of TBT adsorption and desorption in estuarine sediments. Mar Pollut Bull 31:32–43

Langston WJ, Gibbs PE, Livingstone DR (1997) Risk assessment of organotins antifoulings on key benthic organisms of European coastal habitats (BOATS). MAS2-CT-94-0099-Final Report

Langston WJ, Pope ND, Davey M, Langston KM, Hara SCM, Gibbs PE, Pascoe PL (2015) Recovery from TBT pollution in English Channel environments: a problem solved. Mar Pollut Bull 95:551–564

LeDuc DL, Terry N (2005) Phytoremediation of toxic trace elements in soil and water. J Ind Microbiol Biotechnol 32:514–520

Lobinski R, Marczenko Z (1996) Spectrochemical trace analysis for metals and metalloids. Elsevier, Amsterdam

Lorenzana RM, Yeow AY, Colman JT, Chappell LL, Choudhury H (2009) Arsenic in seafood: speciation issues for human health risk assessment. Hum Ecol Risk Assess 15(1):185–200
Losi ME, Frankenberger WT (1997) Environmental chemistry of selenium. Marcel Dekker, New York, pp 515–544
Lovley DR (1995) Advances in agronomy, vol 54. Academic Press, New York, pp 175–231
Lovley DR, Phillips EJP (1992) Appl Environ Microbiol 58(850):65
Lovley DR, Widman PK, Woodward JC, Phillips JP (1993) Appl Environ Microbiol 59:3572
Majumdar A, Upadhyay MK, Ojha M, Biswas R, Dey S, Sarkar S, Moulick D, Niazi NK, Rinklebe J, Huang JH, Roychowdhury T (2024) A critical review on the organo-metal(loid)s pollution in the environment: Distribution, remediation and risk assessment. Sci Total Environ 951:175531. https://doi.org/10.1016/j.scitotenv.2024.175531
Manahan SE (2000) Environmental chemistry.7th edn. CRC Press, Boca Raton
Mania M, Rebeniak M, Szynal T, Mazurek MW, Starska K, Ledzion E, Postupolski J (2015) Total and inorganic arsenic in fish, seafood and seaweeds - exposure assessment. Ann Natl Inst Hyg 66(3):203–210
Mattos Y, Stotza WB, Romeroa MS, Bravob M, Fillmann G, Castro IP (2017) Butyltin contamination in Northern Chilean coast: is there a potential risk for consumers? Sci Total Environ 595(2017):209–217
Maureen CE, Willingham GL (1996) Degradation of antifouling biocides. Biofouling 10(1–3):239–249. https://doi.org/10.1080/08927019609386283
McCarthy JJ, Canziani OF, Leary NA, Dokken DJ, White KS (2001) Climate change 2001: impacts, adaptation, and vulnerability. Cambridge University Press, Cambridge
Meador JP (2000) Predicting the fate and effects of tributyltin in marine systems. In: Ware GW (ed) Reviews of environmental contamination and toxicology, vol 166. Springer, New York, pp 1–48
Meador JP, Krone CA, Dyer DW, Varanasi U (1997) Toxicity of sediment-associated tributyltin to in faunal invertebrates: species comparison and the role of organic carbon. Mar Environ Res 43:219
Mennie D, Craig PJ (1993) Analysis of organometallic compounds in the environment. In: Sigel H, Sigel A (eds) Metal ions in biological systems, vol 29. Marcel Dekker Inc, New York, pp 37–78
Meyer S, Matissek M, Muller SM, Taleshi MS, Ebert F, Francesconi KA, Schwerdtle T (2014) In vitro toxicological characterisation of three arsenic-containing hydrocarbons. Toxicol Res 6:1023–1033
Meyer S, Raber G, Ebert F, Leffers L, Müller SM, Taleshi MS, Schwerdtle T (2015) In vitro toxicological characterisation of arsenic-containing fatty acids and three of their metabolites. Toxicol Res 4(5):1289–1296
Mishra S, Lin Z, Pang S et al (2021) Biosurfactant is a powerful tool for the bioremediation of heavy metals from contaminated soils. J Hazard Mater 418:126253
Molin M, Ulven SM, Meltzer HM, Alexander J (2015) Arsenic in the human food chain, biotransformation and toxicology - review focusing on seafood arsenic. J Trace Elem Med Biol 31:249–259
Moore JW (1990) Inorganic contaminants of surface water. Springer, NewYork
Morcillo Y, Porte C (1998) Monitoring of organotin compounds and their effects in molluscs from the Catalan Coast (NW Mediterranean). Trends Anal Chem 17:109–116
Morcillo Y, Borghi V, Porte C (1997) Survey of organotin compounds in the Wester Mediterranean using molluscs and fish as sentinel organisms. Arch Environ Contam Toxicol 32:198–203
Moreda-Pineiro J, Alonso-Rodríguez E, Moreda-Piñeiro A, Moscoso-Pérez C, Muniategui-Lorenzo S, Lopez-Mahía P, Prada-Rodríguez D, Bermejo-Barrera P (2010) Simultaneous pressurized enzymatic hydrolysis extraction and clean up for arsenic speciation in seafood samples before high-performance liquid chromatography–inductively coupled plasma-mass spectrometry determination. Anal Chim Acta 679(1–2):63–73
Nordstrom S, Beckman L, Nordenson I (1979) Occupational and environmental risks in and around a smelter in northern Sweden. V. Spontaneous abortion among female employees and decreased birth weight in their offspring. Hereditas 90:291–296
Nsengimana H (2007) Speciation of organometallic of tin, lead, and mercury in environmental samples. Ph.D thesis, Faculty of Sciences, University of the Witwatersrand, Johannesburg, p 293

Ohji M, Harinom H, Hayashizaki K, Inoue K, Yusoff M, F., Nishida, S. (2019) Accumulation of organotin compounds on mangroves in coastal ecosystems. J Mar Biol Assoc. https://doi.org/10.1017/S0025315419000262

Okoro HK, Fatoki OS, Adekola FA, Ximba BJ, Snyman RG (2011) Sources, environmental levels and toxicity of organotin in marine environment-a review. Asian J Chem 23(2):473–482

Oppenheimer M et al (2019) Sea level rise and implications for low lying islands, coasts and communities. In: IPCC special report on the ocean and cryosphere in a changing climate. IPCC, Geneva

Pacyna EG, Pacyna JM, Steenhuisen F, Pacyna SW, Steenhuisen JM (2006) Global anthropogenic mercury emission inventory for 2000. Atmos Environ 40(22):4048–4063

Patai S (ed) (1995) The chemistry of organic germanium, tin and lead compounds. Wiley, Chichester

Patocka J (2008) Organic lead toxicology. Acta Med Austriaca 51(4):209–213

Pellerito L, Barbieri, De Stefano R, Scopelliti M, Pellerito C, Fiore T, Triolo F (2002) Toxic effects of organometallic compounds towards marine biota. In: Gianguzza A, Pelizzetti E, Sammartano S (eds) Chemistry of marine water and sediments. Environmental science. Springer, Berlin. https://doi.org/10.1007/978-3-662-04935-8_15

Quintas PY, Arias AH, Alvarez MB et al (2021) Distribution of butyltin compounds in the coastal environment of the Bahía Blanca Estuary. Argentina Arch Environ Contam Toxicol 81:307–323. https://doi.org/10.1007/s00244-021-00871-x

Radojevic M, Harrison RM (1987) Concentrations and pathways of organolead compounds in the environment: a review. Sci Total Environ 50:157

Rudovica V, Bartkevics V (2015) Chemical elements in the muscle tissues of European eel (Anguilla anguilla) from selected lakes in Latvia. Environ Monit Assess 187:1–9

Sakakibara M, Watanabe A, Inoue M et al (2012) Phytoextraction and phytovolatilization of arsenic from As-contaminated soils by Pteris vittata. In: Proceedings of the annual international conference on soils, sediments, water and energy. University of Massachusetts at Amherst, Amherst

Sarradin P-M, Lapaquellerie Y, Astruc A, Latouche C, Astruc M (1995) Long term behaviour and degradation kinetics of tributyltin in a marina sediment. Sci Total Environ 170(1-2):59–70

Scianni C, Georgiades E, Mihaylova R, Tamburri MN (2023) Balancing the consequences of in-water cleaning of biofouling to improve ship efficiency and reduce biosecurity risk. Front Mar Sci 10:1239723. https://doi.org/10.3389/fmars.2023.1239723

Shibata Y, Morita M (1992) Characterization of organic arsenic compounds in bivalves. Appl Organometal Chem 6:343–349

Shiomi K, Kakehashi Y, Yamanaka H, Kikuchi T (1987) Identification of arsenobetaine and a tetra methyl arsonium salt in the clam Meretrix lusoria. Appl Organomet Chem 1(2):177–183

Simon S (2002) Transfer of organotins from water to plants development of a pretreatment process before analysis. Elsevier, Amsterdam, pp 1–18

Smith PI (1998) Chemistry of tin. Blackie Acad and Prof, London

Streets D, Hao J, Wang S, Wu Y (2009) Mercury emissions from coal combustion in China. In: Mason R, Pirrone N (eds) Mercury fate and transport in the global atmosphere. Springer, Boston. https://doi.org/10.1007/978-0-387-93958-2_2

Stromgren T, Bongard T (1987) The effect of Tributyltin oxide on growth of Mytilus edulis. Mar Poll Bull 18(1):30–31

Sunday OA, Alafara AB, Oladele GO (2012) Toxicity and speciation analysis of organotin compounds. Chem Spec Bioavailability 24(4):216–226. https://doi.org/10.3184/095422912X13491962881734

Tamburri MN, Georgiades ET, Scianni C, First MR, Ruiz GM, Junemann CE (2021) Technical considerations for development of policy and approvals for in-water cleaning of ship biofouling. Front Mar Sci 8:804766. https://doi.org/10.3389/fmars.2021.804766

Taylor V, Goodale B, Raab A, Schwerdtle T, Reimer K, Conklin S, Karagas MR, Francesconi KA (2017) Human exposure to organic arsenic species from seafood. Sci Total Environ 580:266–282. https://doi.org/10.1016/j.scitotenv.2016.12.113

Tripathi V, Gaur VK, Dhiman N et al (2020) Characterization and properties of the biosurfactant produced by PAH-degrading bacteria isolated from contaminated oily sludge environment. Environ Sci Pollut Res 27:27268–27278

Ullrich SM, Tanton TW, Abdrashitova SW (2001) Mercury in the aquatic environment: a review of factors affecting methylation. Crit Rev Environ Sci Technol 31:241–293

UNEP (1989) Assessment of organotin compounds as marine pollutants in the Mediterranean. Map Technical Report Series No 33, 1–185

UNEP (2003) Inclusion of the chemicals tetraethyl lead and tetramethyl lead and adoption of the draft decision guidance document. http://www.pic.int/incs/icrc4/s15Add2)/English/ICRC4.15add.2_FINAL.pdf

Varjani SJ (2017) Microbial degradation of petroleum hydrocarbons. Bioresour Technol 223:277–286

Vodopivez C, Curtosia A, Pelletier E, Saint-Louis R, Spairania LU, Hernándeza EA, Zakrajseka A, Geneza A, Mac Cormack WP (2021) Low levels of PAHs and organotin compounds in surface sediment samples from a broad marine area of 25 de Mayo (King George) Island, South Shetland Islands. Sci Total Environ 785:147206

Wang J, Shi L, Zhai L et al (2021) Analysis of the long-term effectiveness of biochar immobilization remediation on heavy metal contaminated soil and the potential environmental factors weakening the remediation effect: a review. Ecotoxicol Environ Saf 207:111261

Wang F, Zhang J, Xu L, Ma A, Zhuang G, Huo S, Zou B, Qian J, Cui YI (2024) Selenium volatilization in plants, microalgae, and microorganisms. Heliyon 10(4):e26023. https://doi.org/10.1016/j.heliyon.2024.e26023

Warford L, James D, Mason C, Lonsdale L, Bersuder P, Blake S, Evans N, Thomas B (2022) A reassessment of TBT action levels for determining the fate of dredged sediments in the United Kingdom. Mar Pollut Bull 176:113439

Wee SY, Aris AZ (2023) Environmental impacts, exposure pathways, and health effects of PFOA and PFOS. Ecotoxicol Environ Saf 267:115663

Weidenhaupt A, Arnold CSR, Muller SB, Haderlein RP, Schwarzenbach C (1997) Sorption of organotin biocides to mineral surfaces. Environ Sci Technol 31:2603–2609

Whaley-Martin KJ, Koch I, Moriarty M, Reimer KJ (2012) Arsenic speciation in blue mussels (Mytilus edulis) along a highly contaminated arsenic gradient. Environ Sci Technol 46(6):3110–3118

Xiaona L, Feng H, Zhenyu W, Baoshan X (2022) Roadmap of environmental health research on emerging contaminants: Inspiration from the studies on engineered nanomaterials. Eco-Environ Health 1(3):181–197

Zhang S, Li P, Li Z-H (2021) Toxicity of organotin compounds and the ecological risk of organic tin with co-existing contaminants in aquatic organisms. Compar Biochem Physiol Part C Toxicol Pharmacol 246:109054. https://doi.org/10.1016/j.cbpc.2021.109054

Zhong H, Wang WX (2008) Effects of sediment composition on inorganic mercury partitioning, speciation and bioavailability in oxic surficial sediments. Environ Pollut 151:222–230

Chapter 4
Mechanisms and Models for OMCs' Chemical Behavior and Ecotoxicology

Abstract Environmental chemistry and biological sciences must work closely together to fully comprehend ecotoxicological impacts. Bioaccumulation lipophilicity is a criterion for the environmental persistence of OMCs. A M–C bond within organometallics might get broken or reformed under ecological conditions. OMC biological activity and environmental risk have gained significant attention. It discusses how temperature, salinity, and pH affect the chemical behavior of OMCs (pH-dependent dual Langmuir model). Pollutant absorption and bioavailability cannot be considered separate chemical or biological issues. This chapter focuses on speciation mechanisms of organometal (loid)s, competitive adsorption-desorption of BTs, and diffusion. Models to calculate TBT release from sediment are displayed. Ecotoxicological aspects of OMCs are deeply discussed with special reference to ecological risk assessment (RQ = PEC/PNEC) to create early warning systems. Influence of environmental factors (pH, salinity) and minerals, organic carbons, and quartz on the ecotoxicity of OMCs are addressed. Organometallic species react with proteins, and the decay of the myelin coating on nerve fibers is a two mode of action for OMC toxicity. Ca^{2+}-dependent and Mg-ATPase activity are two mechanisms highlighting the toxicity of organotin compounds. Human health risk via tolerable average residue level (TARL), toxic unit (TU), cancer risk (CR), risk quotient (RQ), screening level (SL), target hazard quotient (THQ), and probabilistic distribution ($\text{Intake}_{i,k}\,(x, y)$) are evaluated. The biosensing potential of OMCs in aquatic systems is assessed, and two examples of biomarkers of TBT in marine invertebrates (molecular genetic and imposex) are discussed. An overview highlighting the cutoff points of the ecological evaluation technique for OMCs is presented.

Keywords Environmental chemistry · Speciation · Ecological and human risk · Biomarkers

T. O. Said, G. M. El Zokm, *Organometallic Compounds in the Marine Environment*, https://doi.org/10.1007/978-3-031-99189-9_4

4.1 Chemical Behavior of OMCs (Sheds Light on Drawbacks)

Many authors reported that physical, chemical, and biological processes in the environment mediate the environmental chemistry of OMCs (Kucharski et al. 2022). Organic matter and biota aggregation for OMCs revealed strong coefficients of octanol-water [K_{ow}], organic carbon [K_{oc}] partitioning, and weak water solubility (Batista-Andrade et al. 2018). TBT-OH exhibits an acid-dissociation constant (pKa) of 6.3 for the TBT cation in salt water. Additionally, cations of TBT and other OTs are drawn to and adsorbed onto suspended particles with net negatively charged surfaces, like organic matter and clay. The cationic form of TBT and TFT (Bu_3Sn^+ and Ph_3Sn^+) is stable at lower acidity levels; however, the two forms primarily exist as chlorides and hydroxides in an aqueous medium with a pH of 6–8 (ideal seawater) (de Oliveira et al. 2010; Temerdashev et al. 2023). However, OMCs are also present as neutral molecules in aquatic environments. Their bioaccumulation and negatively impacted aquatic systems are relevant to cation form. Many extremely poisonous chemicals have been found in OMCs; this has been linked to the lipophilic characteristics of OMCs, which accelerate the resorption rate. Most organometal[loid]s undergo metabolization to monovalent cations following resorption by losing an organic substituent. Due to their dual lipophilic and hydrophilic characteristics, they easily pass through cell membranes and are transported through bodily fluids. Because highly effective concentrations in target organs can be reached quickly and easily, the primary toxicity is frequently attributed to these monovalent cations (Craig 2023). Among the various toxicities that OMCs (e.g., TBTs) may cause are increased larval mortality (IPCS 1990), inhibited egg development, induced vitellogenin gene expression, reduced hatchling survival, altered population sex ratios, induced DNA damage, increased expression of heat-shock proteins, and female "intersex" condition in certain *Meso gastropoda* species, like the common periwinkle *Littorina littorea*. TBT usually degrades via biological and physicochemical processes, such as biotransformation into less hazardous secondary metabolites by fish, algae, bacteria, fungi, and mammals (Gomez et al. 2024). A few examples of these biological processes are hydrolysis and photodegradation. Up to several years pass before an OMC breakdown happens in the bottom sediments, which is substantially slower than in water. Physicochemical properties of OMCs (i.e., weak water solubility, strong coefficients of octanol-water [K_{ow}], and organic carbon [K_{oc}] partitioning) encourage separation into organic matter and biota aggregation. Several OMCs exploded with water and spontaneously set fire in the air, especially the ones that have small alkyl groups bound to reactive metals. Others are quite dangerous, particularly those containing cadmium and mercury. Some of these substances are easily absorbed via the skin, which exacerbates the issue. Several mechanisms are proposed to discuss OMC toxicity, especially reproductive toxicity (Mandal et al. 2024; Panga and Zhao 2024). The metalloid's potency and dosage both affect how hazardous OMCs are. Though research on the site-specific dose of the metalloid and the potential for biotransformation that might take place during the exposure time is still lacking, it is still rarely addressed in studies. A powerful feature of OMCs is their ability to break down into inorganic molecules, which are essentially nontoxic. Powerful degradation strategies include physicochemical processes (e.g., hydrolysis and photodegradation) and biological processes (biotransformation) illustrated by Gomez et al. (2024).

The chemistry of organometallic compounds has become one of the most important areas of chemistry, and it has erased any boundary that might have existed between inorganic and organic chemistry. Depending upon the elements in the periodic table, organometallic chemistry is mainly classified into main groups: metal, transition metal, lanthanide, and actinide-based organometallics. You can do things with organometallic chemistry that you cannot do using straight-up organic chemistry. The organic compounds contain at least one M–C bond in which the metal is directly attached to the carbon atom, in which there should be a bonding interaction (covalent, ionic, localized/delocalized) (Hosmane 2017; House and House 2016; Aziz et al. 2024). Many authors reported that physical, chemical, and biological processes in the environment mediate the environmental chemistry of OMCs (Kucharski et al. 2022). Organic matter and biota aggregation for OMCs revealed strong coefficients of octanol-water [K_{ow}], organic carbon [K_{oc}] partitioning, and weak water solubility (Batista-Andrade et al. 2018). TBT-OH exhibits an acid-dissociation constant (pKa) of 6.3 for the TBT cation in salt water. Additionally, cations of TBT and other OTs are drawn to and adsorbed onto suspended particles with net negatively charged surfaces, like organic matter and clay. The cationic form of TBT and TFT (Bu_3Sn^+ and Ph_3Sn^+) is stable at lower acidity levels; however, the two forms primarily exist as chlorides and hydroxides in an aqueous medium with a pH of 6–8 (ideal seawater) (WHO 1999; Temerdashev et al. 2023). Although OMCs are also neutral molecules in aquatic environments, their bioaccumulation, leading to a negative impact on marine systems, is relevant to the cationic form. Physicochemical properties of OMCs (i.e., weak water solubility, strong coefficients of octanol-water [K_{ow}], and organic carbon [K_{oc}] partitioning) encourage separation into organic matter and biota aggregation. Several OMCs exploded with water and spontaneously set fire in the air, especially those that have small alkyl groups bound to reactive metals. If organometallic compounds are considered to be those in which a metal is bonded to an organic group, the range of materials is enormous. Such compounds include metal alkyls, metal olefin complexes, GRs, metallocenes, and metal-carbonyl complexes (Wynne 1988; Schaller et al. 2015). It should also be mentioned at the outset that numerous organometallic compounds constitute hazardous materials. Several of them, especially those containing small alkyl groups bonded to reactive metals, are spontaneously flammable in the air and react explosively with water. The most commonly encountered types of problems in organometallic chemistry are illustrated in Fig. 4.1.

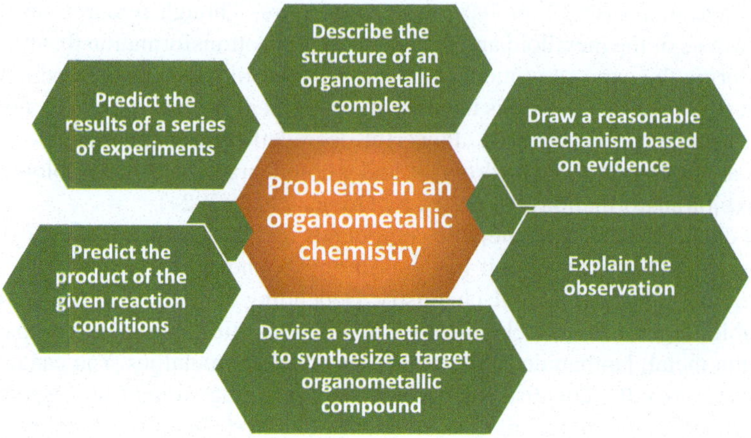

Fig. 4.1 The most commonly encountered types of problems in organometallic chemistry (self-design)

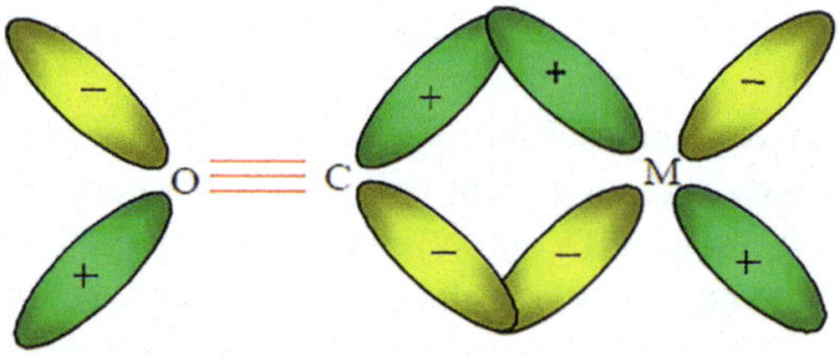

Fig. 4.2 The overlap of a d orbital on a metal with the p* orbital on CO leading to back donation (Elias and Gupta 2010)

4.1.1 Stability of OMCs

OMCs are characterized by stretching bands for bridging CO ligands with metals (Fig. 4.2). Metal carbonyls are reactive compounds, so it is possible to carry out a large number of reactions that lead to useful derivatives. Both the metal and ligand are soft according to the Lewis acid-base definitions.

The oxidation of the organometallic compounds will play a vital role in the stability of the compounds, whereas the organometallic compounds are thermally unstable due to the negative free energy of carbon dioxide, metal oxide, and water production (Fig. 4.2). In addition to that, at lower room temperatures, the organometallic compounds are unstable to oxidation due to the presence of empty orbitals of metal or due to nonbonding electrons (Elias and Gupta 2010).

4.1.2 *Basic Concepts in OMC Chemistry*

4.1.2.1 18-Electron Rule

The structure and bonding present in the OMCs are studied based on the 18-electron rule, and this is the combination of the ability of the ligand for pi acceptor or sigma donor and the back bonding and bonding nature of the ligand. In addition to that, the stability of the metal complexes is predicted and characterized by the 18-electron rule. A total of 18 electrons are present in the valence shell of MT, whereas a total of 10 electrons are in the d orbital (2 each from five d orbitals), 6 electrons from the p orbital (2 each from 3 p orbitals), and 2 electrons from the 1s orbital. The combination of these atomic orbitals ($5d^{+3}p^{+1}s$) results in 9 molecular orbitals. These 9 molecular orbitals are metal-ligand nonbonding or bonding orbitals, and there might be a few higher-energy antibonding molecular orbitals. These 9 molecular orbitals are filled by electrons originating from ligands or metals (Elias and Gupta 2010). Thermodynamically stable transition metal organometallic compounds are formed when the sum of the metal d electrons and the electrons conventionally considered as being supplied by the surrounding ligands equals 18. In this way, the metal attains the electronic configuration of the next higher noble gas in general ns^2 $(n - 1)$ d^{10} np^6.

Example: $Fe(C_5H_5)_2$.

Electronic configuration: $3s^23p^63d^64s^2$.

By neutral counting method: for Fe 8 electrons, for $(C_5H_5)_2$ electrons:

$Fe(C_5H_5)_2$: 8 + 2(5) = 8 + 10 = 18 valence electrons.

4.1.2.2 Back Bonding or Back Donation

When electrons move from one atomic orbital to another antibonding orbital π^* of a ligand or atom, the process relieves the metal with an excess negative charge.

Example: Ni $(CO)_4$.

4.1.2.3 Hapticity *ηx*

The term hapticity is defined as the total number of donor atoms of a ligand that are coordinated (attached) to the central metal atom (it is also described as a contiguous group of atoms of a ligand, which are coordinated to the central metal atom). Hapticity is denoted by the Greek word *ηx*, where *x is* number of donor atoms attached to the metal. Ligands are atoms or groups of atoms, ions, or molecules capable of donating a pair of electrons to the metal atom and are called ligands. The hapticity of ligands ranges from 1 to 8, that is, monohapto ligands to octohapto ligands. Discrimination of organometallics based on the M–C bond is illustrated in Fig. 4.3 (Cornelis, 2007).

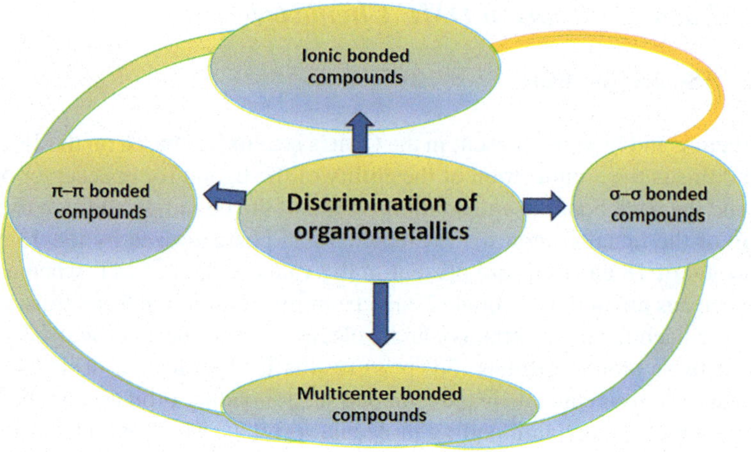

Fig. 4.3 Discrimination of organometallics based on the metal-carbon bond based on Cornelis (2007)

4.1.3 Methylation, Bioavailability, and Hydrolysis

The hydrolysis of organometallic compounds will also affect the stability of compounds: (1) hydrolysis involves the reaction of nucleophilic attack of water, (2) the polarity of the M-C bond is the rate-determining step, and (3) the higher the polarity of the M–C bond, the faster the rate of hydrolysis.

4.1.3.1 Case Study (Hydrolytic Reactions)

The hydrolytic reactions of butyltins (BTs) in water with the acidity constant (pKa) (Fang et al. 2017) (Eqs. (4.1)–(4.6)).

$$TBT^{+} + H_2O \rightleftarrows TBTOH + H^{+} \quad 6.3 \pm 0.1 \tag{4.1}$$

$$DBT^{2+} + H_2O \rightleftarrows DBT(OH)^{+} + H^{+} \quad 3.0 \pm 0.1 \tag{4.2}$$

$$DBT(OH)^{+} + H_2O \rightleftarrows DBT(OH)2 + H^{+} \quad 5.1 \pm 0.2 \tag{4.3}$$

$$MBT^{3+} + H_2O \rightleftarrows MBT(OH)^{2+} + H^{+} \quad 2.3 \pm 0.1 \tag{4.4}$$

$$MBT^{2+} + H_2O \rightleftarrows MBT(OH)_2^{\ +} + H^{+} \quad 3.5 \pm 0.2 \tag{4.5}$$

$$MBT(OH)_2^{\ +} + H_2O \rightleftarrows MBT(OH)3 + H^{+} \quad 5.9 \pm 0.1 \tag{4.6}$$

(Weidenhaupt et al. 1997), BCs (Fang et al. 2017), organic matters, natural soils, and sediments (Burton et al. 2004).

4.2 Mathematical Models Describing OMCs' Chemistry

4.2.1 Competitive Adsorption and Desorption of BTs

Sediments are complex and heterogeneous matrices containing clays, sand (quartz), natural organic matter (NOMs), and BCs (black carbon). The total adsorbed TBT by the sediment, q_{total}, is given by Eq. (4.13):

$$q_{\text{total}} = \text{Cw} f_k K_k + C_w f_s K_s + C_w f_{\text{OM}} \text{K}_{\text{OM}} + f_{\text{BC}} C_w K_{\text{BC}} \quad (4.13)$$

where f_k f_s, f_{NOM}, and f_{BC} represent the fractions of kaolinite, sand, OM, and BC, respectively. K_K, K_s, K_{NOM}, and K_{BC} represent the distribution coefficients of TBT with kaolinite, sand, OM, and BC, respectively. C_w represents the concentration of TBT in the aqueous phase. Competition for adsorption sites between the positively charged BTs and background ions such as Ca^{2+} and Mg^{2+} may seriously affect the adsorption of BTs to adsorbents. Weidenhaupt et al. (1997) studied the adsorption of TBT to minerals at pH 4 in the presence of different cations and showed that the adsorption of TBT significantly declined when the monovalent cations' electrolyte (e.g., Na^+ and K^+) was replaced by bivalent cations (e.g., Mg^{2+}, Ca^{2+}, and Ba^{2+}).

The desorption behavior of BTs from environmental matrices can significantly alter the bioavailability and toxicity of BTs in the environment. In contrast, only a few studies have been done on the desorption of BTs (Burton et al. 2004; Huang and Matzner 2004). It suggests that the desorption of BTs follows biphasic kinetics with an initial rapid desorption and a later slower desorption (Ma et al. 2000). The desorption rate of BTs from sediment is highly dependent on the property of BTs, with an order of TBT > DBT > MBT at pH 8 and 4 (Huang and Matzner 2004). Moreover, BT desorption from environmental matrices highly depends on environmental factors (e.g., pH and salinity) and the properties of matrices, which are specified below.

4.2.2 Models Controlling TBT Diffusion

4.2.2.1 Model 1

The vertical diffusive flux of TBT (J) was calculated with $\delta C/\delta z$ values following the equation of diffusion given by Fick's law:

$$J(\text{Vertical diffusive flux}) = -\Phi D_s (\delta C / \delta z) \quad (4.14)$$

where J is the vertical flux, Φ is the sediment porosity, $\delta C/\delta z$ is the concentration gradient, and D_s is the bulk sediment molecular diffusion coefficient.

The bulk sediment molecular diffusion coefficient (D_s) was given by

$$D_s = D_0 / \theta^2 \tag{4.15}$$

where θ is the tortuosity and D_0 is the diffusion coefficient in water.

The diffusion coefficient of OMCs can be estimated according to Eq. (4.15) (D_0 diffusion coefficient in free solution (10^{-6} cm^2 s^{-1}), M molecular mass (g/mol), valid for M = 16–200,000 g/mol) (Hamer and Karius 2005):

$$\log D_0 \left(25°\mathrm{C}\right) = 1.72 - 0.39.\log M \tag{4.16}$$

In aquatic environments, the main TBT species will be TBTOH (molecular weight = 307 g/mol). D_0 can be calculated for TBTOH to 177 cm^2/a. Diffusion is dependent mainly on dynamic viscosity (η [cp]), which changes with temperature (T [K]), salinity (S [PSU]), and pressure (P[bar]). The diffusion coefficient for estuarine water (D_w [cm^2/a]) was then calculated based on Eq. (4.17) (T_0 = 298.15 K, P_0 = 1 bar, S_0 = 0.0 PSU) after Otto (1996):

$$D_w\left(T, P, S\right) = D_0\left(25°\mathrm{C}\right)\frac{\left(T_0, P_0, S_0\right)}{\left(T, P, S\right)}.\frac{T}{298.15} \tag{4.17}$$

4.2.2.2 Model 2

The mean release of TBT from the sediment in the modeled area was calculated by subtracting the total TBT content within the model area after a model run from the total TBT content at the starting point. Due to the high Kd-values, the amount of TBT in the pore water was neglected, and only the TBT adsorbed to the solid phase was taken into account. The amount of TBT in the sediment at a given time (TBT (ti) (mol/m^2)) was calculated with Eq. (4.18) (Φ = porosity (%), dx = discretization (cm), Ci = TBT-concentration in each cell (mol/dm^3 pore water):

$$\mathrm{TBT}\left(\mathrm{ti}\right) = \Phi.10.dx\sum\nolimits_1^i \mathrm{Ci}\left(\mathrm{TBT}\right) \tag{4.18}$$

The annual mean TBT-flux was calculated with Eq. (4.18) (J_{Sed} = TBT-flux (mol m^{-2} a^{-1}), C_{t1} = concentration at t_1 (mol m^{-2}), C_{t2} = concentration at t_2 (mol m^{-2}), t (a):

$$J_{\mathrm{Sed}} = \frac{C_{t2} - C_{t1}}{t_2 - t_1} \tag{4.19}$$

The removal efficiency by the conventional processes for OTs removal varied widely among the aquatic ecosystems. It is calculated as follows (Eq. (4.20)):

$$\mathrm{RE}(\%) = \frac{\text{Concentrations in influents} - \text{Concentrations in effluents}}{\text{Concentrations in influent}} \tag{4.20}$$

4.2.3 Model Equations Describing Bioconcentration of Organotin

Bioaccumulation lipophilicity is a criterion for the environmental persistence of organotins. Among the organotins, TBT is considered to be an important pollutant because of its extreme toxicity to several organisms and because it tends to bioaccumulate. Bacteria have been reported to display a remarkable ability to accumulate TBT. Marine bivalves accumulate significant amounts of TBT (up to 5 μg g^{-1}). But fish and crustaceans accumulate much lower amounts with efficient enzymatic mechanisms to degrade TBT (Laughlin Jr. et al. 1985). Absorption in mice is also low, and TBT is mainly excreted unchanged via the feces. Mammals and birds accumulate high levels of butyltins in their organs and tissues. In mammalian species, TBT compounds may be metabolized to DBT and related metabolites. An undetermined amount of this compound remains in the fat, liver, and kidneys. Other researchers have undertaken studies to evaluate the bioaccumulation of organotin.

Bioconcentration can be described and modeled based on compartments that are idealized stores for incorporated chemicals. The simplest model is a one-compartment model, where uptake kinetics and bioconcentration can be described by a first-order process according to the following Eqs. (4.21)–(4.24):

$$\frac{dc_b}{dt} = K_1 \cdot C_W - K_2 \cdot C_b \tag{4.21}$$

C_W concentration in water (ng/ml), C_b concentration in organism (ng/g), k_1 uptake rate constant (ml/g × h), k_2 elimination rate constant (h′-1 t: Time (h)).

Integration gives

$$c_b = C_W \frac{K_1}{K_2}\left(1 - e^{-k2t}\right) + C_{b(t=0)}\, e^{-k2t} \tag{4.22}$$

The second term becomes zero and can be deleted because $C_{b(t=0)} = 0$.

Division by C_w gives

$$\frac{c_b}{c_w} = \frac{K_1}{K_2}\left(1 - e^{-k2t}\right) \tag{4.23}$$

At steady state, BCF can be defined as

$$\mathrm{BCF} = \frac{K_1}{K_2} = \frac{C_{b(t=0)}}{C_{w(t=0)}} = \frac{\text{ng organotin / } g \text{ wet weightoranism}}{\text{ng organotin / ml water}} \tag{4.24}$$

If no steady state is reached, BCF is calculated by k_1/k_2 or by c_b/c_w, at the end of the accumulation process. BAF can be classified as follows: BAF < 1000, no accumulation probability; 1000 < BAF < 5000, bioaccumulative organism; BAF > 5000, extremely accumulative organism (Ali et al. 2022).

4.3 Ecotoxicological Aspects of OMCs

Generally, toxicity is defined as "an adverse change from normality, which may be irreversible." Adverse changes "affect the well-being of the organism, either temporarily or permanently," whereas normality is considered "in statistical terms of the normal distribution" (Woolley 2008). Ecotoxicology is an interdisciplinary field that combines environmental chemistry, biochemistry, toxicology, and ecology. The overall goal of the ecotoxicological study is to offer broad principles for assessing the possible harm that contaminants may pose. Descriptive studies continue to constitute the majority of ecotoxicological research on chemical impacts, despite significant advancements over the past 10 years. As a result, there is now a wealth of important scientific information on the impacts of certain contaminants on certain species that are being utilized in regulatory contexts. Thus far, there has been a lot less focus on the creation of broad notions that enable evaluations of impacts based on this more theoretically based approach to ecotoxicology. There have been two developments during the past 10 years. Understanding the underlying molecular processes and patterns of toxicity at various levels of biological structure is a target.

Many extremely poisonous chemicals have been found in OMCs; this has been linked to the lipophilic characteristics of OMCs, which accelerate the resorption rate. Organometallic compounds can impose toxic effects via various mechanisms, mainly via the coordination of organometallic species to proteins, as well as via the decay of the myelin coating on nerve fibers. The presence of lipophilic groups facilitating the transport via lipid membranes is considered one of the reasons for the higher toxicity of organometallic compounds, in comparison to inorganic compounds. Arsenic is an exception to this general rule, while mercury, lead, and tin do not. Most organometal[loid]s undergo metabolization to monovalent cations following resorption by losing an organic substituent. Due to their dual lipophilic and hydrophilic characteristics, these cations can easily pass through cell membranes and are transported through bodily fluids with efficiency. Because highly effective concentrations in target organs can be reached quickly and easily, the primary toxicity is frequently attributed to these monovalent cations (Craig and Jenkins 2004). Among the various toxicities that OMCs (e.g., TBTs) may cause are increased larval mortality (IPCS 1990), inhibited egg development, induced vitellogenin gene expression, reduced hatchling survival (Leung et al. 2006), altered population sex ratios (Sousa et al. 2005), induced DNA damage, increased expression of heat-shock proteins, and female "intersex" condition in certain Meso gastropoda species, like the common periwinkle *Littorina* (Beyer et al. 2022). OMCs are quite dangerous, particularly those containing cadmium, tin, lead, and mercury. Some of these substances are easily absorbed via the skin, which exacerbates the issue. After resorption, the majority of organometal(loid)s lose an organic substituent during metabolization, which results in monovalent ions. It often travels through cell membranes, and these monovalent ions are blamed for its toxicity. The evolutionary alteration of OMCs, which completely alters their molecular identity, significantly impacts their significance (typically referring to their transfer and toxic adverse effects). The metal[loid]'s potency and dosage both affect how hazardous OMCs are. Though research on the site-specific dose of the metal[loid] and the potential for biotransformation that might take place during the exposure time is still lacking, it is still rarely addressed in studies. Many extremely poisonous chemicals have been found in OMCs; this has been linked to the lipophilic characteristics of OMCs, which accelerate the rate of resorption. Most organometal[loid]s undergo metabolization to monovalent cations following resorption by losing an organic substituent. Due to their dual lipophilic and hydrophilic characteristics, these cations can easily pass through cell membranes and are transported through bodily fluids with efficiency. Because highly effective concentrations in target organs can be reached quickly and easily, the primary toxicity is frequently attributed to these monovalent cations (Craig and Jenkins 2004). Among the various toxicities that OMCs (e.g., TBTs) may cause are increased larval mortality (IPCS 1990), inhibited egg development, induced vitellogenin gene expression, reduced hatchling survival (Leung et al. 2006), altered population sex ratios (Sousa et al. 2005), induced DNA damage, increased expression of heat-shock proteins, and female "intersex" condition in certain Mesogastropod species, like the common periwinkle *Littorina littorea* (Beyer et al. 2022).

TBT can be leached from the microplastic present in the environment. As a result, TBT is still constantly emitted into the environment from numerous anthropogenic sources (Çetintürk and Ünlü 2022). Several mechanisms are proposed to discuss OMC toxicity, especially reproductive toxicity (Mandal et al. 2024; Panga and Zhao 2024). The toxicity of the OMCs is influenced more by the alkyl substitutes than the anionic substitutes. Quantitative research into the modalities of hazardous action requires knowledge of the effective concentrations in the organism or at the target site(s), which is crucial for connecting a chemical's destiny and effects. Organometallics are widely employed in both industry and academia. Even though toxicology has advanced quickly in recent years, there are still several widely held "beliefs" that are prevalent in the OMC community despite their historical foundation. These approaches must be dissected in vitro or utilizing whole-organism systems to be able to create biomarkers and biosensors with particular applications that can serve as early warning systems. Mode-of-action (MOA)-based techniques, which integrate chemical and biological analytical methods, ought to advance our knowledge of ecotoxicological consequences and be used in risk assessment going forward. Ecotoxicology, which covers all tiers of biological structure from individual organisms to ecosystems, is the study of how toxic compounds affect living things. Ecotoxicology is an interdisciplinary field that combines environmental chemistry, biochemistry, toxicology, and ecology (Fent 2006).

4.3.1 The Overall Goal of the Ecotoxicological Studies

Calculated half-lives recorded 8.80 as a result of minimal degradation rates in sediments, based on first-order TBT degradation rates in sediment. Descriptive studies continue to constitute the majority of ecotoxicological research on chemical impacts, despite significant advancements over the past 10 years. As a result, there is now a wealth of important scientific information on the impacts of certain contaminants on certain species that are being utilized in regulatory contexts. Thus far, there has been a lot less focus on the creation of broad notions that enable evaluations of impacts based on this more theoretically based approach to ecotoxicology; there have been two developments during the past 10 years. The goal of the other trend is to comprehend the intricate relationships and feedback processes in ecosystems that are impacted by pollution. These two methods strengthen one another and will ultimately advance the target site or sites to evaluate the detrimental impacts of pollutants on adsorption, desorption, degradation, partitioning, hydrolysis, aggregation, and bioaccumulation, which are controlling processes for OMCs' chemical behavior. Formal monopositive cations or species that are produced by losing one organic group from the neutral, completely saturated, organometallic, such as R_3Pb, R_3Sn, and CH_3Hg, often have the most hazardous effects. It should be mentioned that organisms often convert saturated organometallics (R_4Sn, R_4Pb, etc.) to R_3M, etc., which is where their poisonous effects originate. When alkyl groups are joined to metals, they are more hazardous than aryl groups. Although the most hazardous alkyl groups (affixed to a common metal) differ from one organism to another, methyl, ethyl, and propyl groups are often the most hazardous. The trialkyl series has the highest level of tin compound toxicity among the organotin cations. Higher alkyl groups are more harmful to bacteria, invertebrates, and fish, whereas longer-chain alkyl compounds are less dangerous to mammals. Methyl and ethyl groups in trialkyl tins are the most poisonous to mammals. The reduction of the myelin sheath of nerve fibers is the other primary outcome of the introduction of OMC into vertebrates. Edema and water buildup in the central nervous system can also happen. Coordination to nonenzyme sites (such as thiols, protein histidine residues, hemoglobin, cytochrome P_{450}, and brain receptors) results in additional harmful consequences. OMCs may have an impact on an organism's immune system, hematopoietic bone marrow, and vital trace metal systems. Basic disruptions to DNA, protein synthesis, mutagenicity, and genotoxicity have been reported. Consequently, millions of OMC derivatives can be synthesized when accounting for all elements less negative than carbon and the total number of organic compounds currently in existence. Some of these are widely used in organic syntheses, while others may be used as pesticides, fire retardants, wood preservatives, and antifouling agents, and other applications are used extensively in industry, agriculture, and research labs; nonetheless, their toxicity poses a risk to the environment. However, nothing is known about how the substituents attached to the core atom M affect toxicity and other biological characteristics of BP. Alkyl derivatives of metals are more toxic than aryl ones, with organic derivatives generally being more harmful than the parent inorganic metal. Moreover, the coordination atoms in the attached biological molecule determine the toxicity mechanism. Egorochkin et al. (2013) used correlation analysis to examine data from the literature on substituent influence on the toxicity (lethal dosages, LD_{50} and LDL_0) of compounds containing mercury, boron, silicon, germanium, tin, lead, phosphorus, arsenic, and platinum (27 series). Several biological properties (BPs) have been compared, including mutagenicity (4 series), antitumor, fungicidal, and insecticidal actions. The BP was initially found to be dependent on the polarizability of substituents, which is defined by the $\sigma\alpha$ constants, in addition to the inductive, resonance, and steric effects. The donor-acceptor interaction between the OM molecule and the biological target causes an excess charge to emerge on the M atom, which is the cause of the polarizability effect. The type of series determines whether certain effects are present or absent. Most of the time, only when the polarizability effect is taken into account can the effects of the substituents on BP be accurately explained. Understanding substituent effects helps one better comprehend the variables affecting OMCs' biological activity. For the first time, they discovered that inductive, resonance, and steric effects not only cause the toxicity (lethal dose LD_{50}) of compounds containing silicon, germanium, tin, phosphorus, boron, and mercury, but they also regulate the polarizability of substituents bonded to the central atom M in organometallics. The extra charge on M that results from the complexation of organometallics with a biological target is the cause of the polarizability effect. According to some ecotoxicologists, TBT is the most dangerous man-made chemical ever purposefully put into the environment in significant amounts. Over more than 50 years of research, thousands of studies about TBT's environmental fate and effects have been reported. The toxicity of the butyl- and phenyltin derivatives in live organisms depends on dose, time of exposure, the type and specificity of a given species, the age and resistance of an individual, and the physiological processes occurring in a given organism. Note that mollusks and snails accumulate the most organotin and are the main source of pollutants delivered to the remaining links in the food chain. Next are the crustaceans and fish. Lipophilicity of organotin results in their propensity to persist and bioaccumulate in the sequential links of the food chain. This may lead to the spread of toxic substances in the food web and, in particular cases, to the poisoning of specific organisms. Several organotins are strongly immunosuppressive, display developmental and reproductive effects, and are neurotoxic. Acute and chronic toxic effects of OTC on microorganisms are reported in Tables 4.1 and 4.2. The toxicity of alkyl lead diminishes in the following sequence:

$$R_4Pb > R_3Pb^+ > R_2Pb^{2+}\left(R = C_2H_5 > CH_3\right)$$

Table 4.1 Comparison of acute toxicities of metal salts and organometallic compounds (Egorova and Ananikov 2017; TOXNET 2017)

Entry	Compound	Mouse	Rat
	Organometallic compounds		
1	Monobutyltin trichloride	1400	2140
2	Dibutyltin dichloride	35	150
3	Tributyltin chloride	117	129
4	Tetramethyllead		105
5	Disodium methyl arsonate hexahydrate		821
	Metal salts		
6	$CuCl_2$	190	140-584
7	$ZnCl_2$	350	350-1100
8	$PdCl_2$	>1000	200-2704
9	$PtCl_2$		3423
10	$PtCl_4$		240-276
11	$SnCl_2$	1200	700

Rating	Common description	LD_{50} (single oral dose for rats, mg kg^{-1})
1	Extremely toxic	≤1
2	Highly toxic	1-50
3	Moderately toxic	50-500
4	Slightly toxic	500-5000
5	Practically non-toxic	5000-15000
6	Relatively harmless	>15000

[a] LD50, single oral dose tested in rats (Carson 2002)

Table 4.2 Acute toxicity of organotin compounds

Compound	Mouse	Rat
Monobutyltin trichloride	1400	2140
Dibutyltin dichloride	35	150
Tributyltin chloride	117	129
Tributyltin oxide	85	122
Triphenyltin acetate	81	136
Triphenyltin chloride		80
Triphenyltin hydroxide	209	171
Azocyclotin		209
Tricyclohexyltin hydroxide (cyhexatin)	710	540
Fenbutatin oxide	1450	4400

[a] Oral LD50, in mg kg^{-1}. Wiley-VCH Verlag GmbH & Co. KGaA (Egorova and Ananikov 2017)

However, the ionic species are more persistent in the environment. Similarly, methylated lead species are less toxic than the corresponding ethylated compounds but more stable and volatile and of longer half-lives. Although there are indications that biologically mediated mechanisms can methylate lead, the natural background level of methyl lead species is negligible due to the reportedly low efficiency of such processes. Vehicle exhaust fumes are undoubtedly the primary source of alkyl lead species in the environment. Nevertheless, these findings highlight the bioavailability of lead in the environment and the necessity of tracking specific alkylated species to investigate the chemical cycling of these substances and evaluate the risks of exposure (Abadin and Pohl 2010).

4.3.2 Mechanisms Describing Butyltin Toxicity

The potential toxicity of butyl tins depends on their high affinity for the cell membranes. Their ability to cross the biological membranes permits them to penetrate easily into cells and their organelles and interact with several intracellular targets. The biocidal properties are influenced strongly by the *R* (1–3) Sn cation and are relatively independent of the anionic radical-associated anions (Champ and Seligman 1996). Mechanisms underlying acute and chronic toxicity usually differ; however, these two types of toxicities cannot always be easily distinguished from each other, and in some cases, acute toxicity can lead to chronic toxicity. It should be noted that the possibility of acute toxicity manifestation depends on the substance dose: even beneficial and safe compounds can cause acute toxicity when a sufficiently large dose is used. Still, the dose increase does not necessarily mean the development of chronic toxicity (Dom et al. 2012).

Most toxic effects of butyl tin are irreversible, are dose- and time dependent, and can be divided into Ca^{2+}-independent and Ca^{2+}-dependent mechanisms, but they are strictly connected.

4.3.2.1 Ca^{2+}-Dependent Model

Both planned cell death and toxic cell death depend on calcium ions. One of the first steps in the development of cell injury has been identified as the disruption of intracellular Ca^{2+} homeostasis brought on by either increased Ca^{2+} influx or suppression of Ca^{2+} extrusion. Organotin compounds induce alterations in the calcium homeostasis of the cells, increasing concentrations of internal calcium ions (Laughlin Jr. et al. 1985; Orrenius et al. 1992). This effect is a multifactorial process, involving a release of Ca^{2+} from intracellular stores due to an increase in the membrane Ca^{2+} permeability of cell compartments. A decrease in Ca^{2+}-ATPase activity represents the main Ca^{2+} extrusion system (Cima et al. 2002). Recently, a direct interaction between butyltins and calmodulin, probably hydrophobic (between the aliphatic chains of butyltins and hydrophobic pouches of Ca^{2+}-activated calmodulin), has been proposed. It follows a saturation pattern which leads to the formation of a complex, preventing the regulative activity of calmodulin on the membrane-bound calmodulin-dependent Ca^{2+}-ATPase (Cima et al. 2002). Results from several biochemical studies (Chow et al. 1992; Fent 1996a, b; Bouchard et al. 1999) indicate that the immunotoxicity of butyl tins in mammals is caused by an alteration of Ca^{2+} homeostasis, resulting in an alteration of cellular structure and functions. TBT-increased $[Ca^{2+}]_i$ occurs by multiple mechanisms (Fig. 4.5), including the release of the inositol triphosphate-sensitive intracellular Ca^{2+} stores, stimulation of a

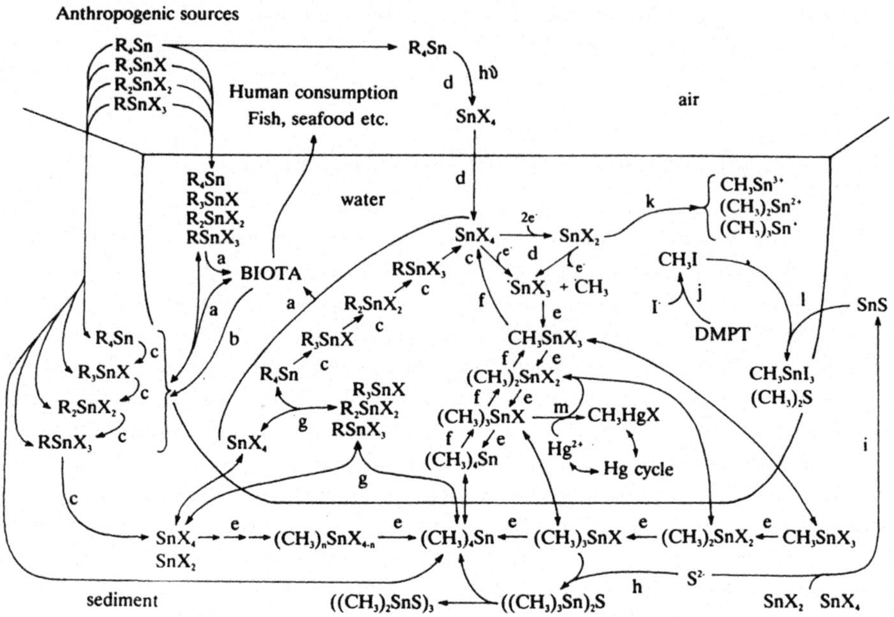

Fig. 4.5 Possible biogeochemical cycle of organotin compounds (Gadd, 2000). The routes listed are (**a**) bioaccumulation, (**b**) deposition or release from biota, (**c**) biotic or abiotic degradation, (**d**) photolytic degradation resulting in the production of free radicals, (**e**) biomethylation, (**f**) demethylation, (**g**) disproportionation reactions, (**h**) sulfide-mediated disproportionation reactions, (**i**) SnS formation, (**j**) formation of methyl iodide by reaction of dimethyl β-proprotein (DMPT) with aqueous iodide, (**k**) CH3I methylation of SnX2, (**l**) oxidative methylation of SnS by CH3I with formation of methylation triiodide, and (**m**) transmethylation reaction between organotins and mercury. Abbreviations: *R* organic moiety; *X* anion (Gadd, 2000)

Ca^{2+}-entry pathway, and inhibition of Ca^{2+} efflux by the plasma membrane Ca^{2+}-ATPase. Using a simple structure-activity approach, DBT appears to be a better competitor for calcium sites than TBT because the ionic form of DBT is a double-charged cation like Ca^{2+}, and DBT is a much smaller moiety than TBT, which brings only one positive charge. The second factor to be considered is the possible partial retention of TBT at the surface or within the plasma membrane of hemocytes. The ability of TBT to induce major modifications in plasma membranes was recently established using anisotropy measurements of fluorescent probes embedded in the phospholipidic layer of isolated trout erythrocyte membranes (Fig. 4.6) (Orrenius et al. 1992).

4.3.2.2 Inhibition of the Mitochondrial Mg-ATPase Activity Model

According to the kinetic approach, evidence that both TBT and DBT in vitro inhibit the Mg-ATPase in mussel digestive gland mitochondria by a different mechanism (Nesci et al. 2011). DBT displays higher efficiency than TBT (IC50 = 0.32 μM for TBT vs. 0.19 μM for DBT) (Fig. 4.7).

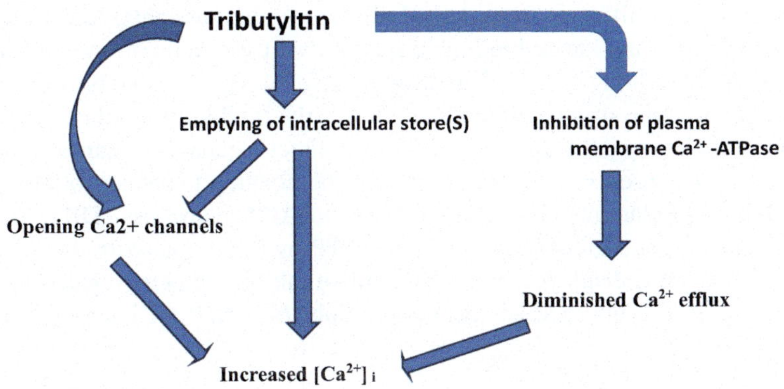

Fig. 4.6 Multiple mechanisms of $[Ca^{2+}]_i$ elevation by the immunotoxicity TBT (Orrenius et al. 1992)

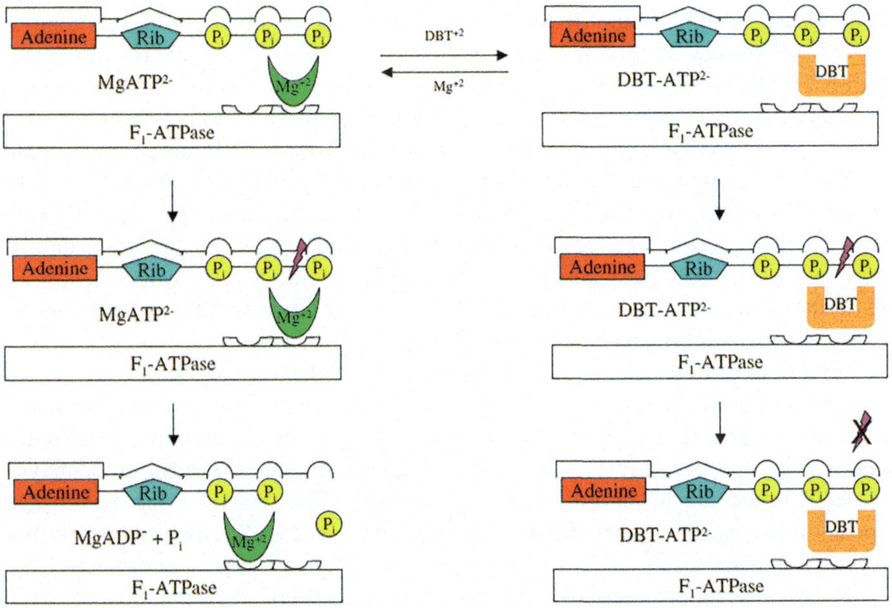

Fig. 4.7 Hypothesized model of F1 inhibitory mechanism by DBT. In the catalytic site, Mg^{2+} interacts with the b and c phosphate groups of ATP and facilitates nucleophilic attack by water molecules, leading to ATP hydrolysis (left). DBT (right) interacts with phosphate groups, displaces Mg^{2+}, and prevents ATP hydrolysis (Nesci et al. 2011)

It seems likely that TBT, due to its lipophilicity and capability of reaching transmembrane regions (Chicano et al. 2002), may be inserted in the H^+ translocation region of mussel mitochondrial Mg-ATPase F0. Powers and Beavis (1991) hinted that the TBT binding site on the F0F1 ATPase was accessible from the lipid bilayer.

The TBT site was different from that of oligomycin in mammals. Accordingly, the TBT localization and steric hindrance may prevent oligomycin and DCCD from reaching their binding sites. TBT hydrophobicity may favor its stabilization within nonpolar membrane structures. On these bases, the inhibition of ATP hydrolysis of the F_0F_1 complex by TBT may be due to an indirect mechanism, namely, the block of proton flux caused by TBT may prevent the counterclockwise rotation of this "splendid molecular machine," which leads to ATP hydrolysis. DBT, shown to behave as a more efficient inhibitor of the ATP hydrolytic activity of the enzyme complex than TBT, displayed an uncompetitive inhibition, being unable to bind to the enzyme in the absence of the natural substrate ATP (K'i tends to infinite) (Boyer 1997).

4.4 Case Study 1: Organotin Toxicity

The toxicity of the organotin is influenced more by the alkyl substitutes than the anionic substitutes. Progressive introduction of organic groups to the tin atom in any number of the $RnSnX_{4-n}$ series produces maximal biological activity against all species when $n = 3$ (Dubey and Roy 2003; Antizar-Ladislao 2008). Tributyltin oxide (TBTO) and tributyltin chloride have been normally used in laboratory experiments to investigate organotin toxicity. In the aquatic environment, TBT is quickly removed from the water column and adheres to bed sediments because TBT has a high specific gravity near 1.2 kg/L at 20 °C (Landmeyer et al. 2004), low solubility less than 10 mg/L at 20 °C and pH 7.0 (Fent 1996a), and log K_{ow} values near 4.4 at pH 8 (Meador 1997, 2000; Meador et al. 1997). Additionally, TBT is ionizable and exhibits a pK_a acidity constant of 6.25 (Meador 2000). The affinity of organotin for adsorption to sediments is positively correlated to the extent of organo-substitution on the tin (Landmeyer et al. 2004). TBT is an endocrine-disrupting compound with the capacity to influence biological fitness (e.g., growth, development, and reproduction) in sensitive species groups. In this regard, the focus will be on identifying molecular mechanisms and the modes of actions (MoA) that TBT has on sensitive species and how these mechanisms may be relevant in the toxicity progression, also for other marine species.

The toxicological pattern of organotins is very complex. The biological effects of the substances depend on the nature, number, and type of organic groups R (alkyl or aryl group) bound to the Sn cation. Inorganic anion X plays a secondary role. Organotin toxicity to microorganisms decreases in the order $R_3Sn\ X > R_2Sn\ X_2 > RSn\ X_3$, with R_4Sn compounds being of low toxicity. In general, tripropyl-, tributyl-, and triphenyltins are the most toxic, with an association with cell membranes being an important prerequisite for toxicity. However, Cima et al. (2002) observed a significantly more effective inhibition of phagocytosis by DBT than TBT and also found a more important and significant reduction of Ca^{2+} adenosine triphosphatase (ATPase) activity by DBT than TBT. Inhibition of Ca^{2+}-ATPase by TBT and DBT might result from competition for enzyme-active sites usually occupied by Ca^{2+}. The inhibition of microbial processes by TBT has been recorded for all major groups, with the main interactions occurring at cellular membranes and chloroplasts or the mitochondria in eukaryotes (Sunday et al. 2012).

4.4.1 Toxicity to Marine Invertebrates

TBT causes impairments in growth and development and induces reproductive failures, shell anomalies, and gel formation. It also causes chambering and high mortality, disturbs the energy metabolism of bivalves, and inhibits the activity of many enzymes; these effects reduce the survival of many species (Haggera et al. 2005). TBT, as early as the 1970s, was known to be very toxic to many aquatic organisms. The high toxicity of TBT is attributed to its effects on mitochondrial function. The embryonic and larval stages of marine invertebrates are less tolerant to toxicants than adults, and this difference has been used to assess the biological quality of marine water and sediments (Fent and Stegeman 1991).

4.5 Organoselenium Toxicity in Mammals

PhSeSePh inhibited mammalian δALA-D with a relatively high potency by the fact that the mammalian enzyme possesses two cysteine residues in spatial proximity at the active site of the enzyme. The PhSeH formed after the reaction with thiols is oxidized back to PhSeSePh by atmospheric O_2. The regeneration of PhSeSePh after reaction with O_2 helps to explain the previous observation that the inhibitory potency of PhSeSePh toward rat δ-ALA-D decreases considerably in an anaerobic atmosphere (Farina et al. 2002). The inhibitory effect of PhSeO2H on mammalian δ-ALA-D can be mediated by direct oxidation of Cys*-SH and Cys**-SH. Oxidation of other less reactive cysteinyl residues other than those shown in the scheme cannot be ruled out (Fig. 4.8).

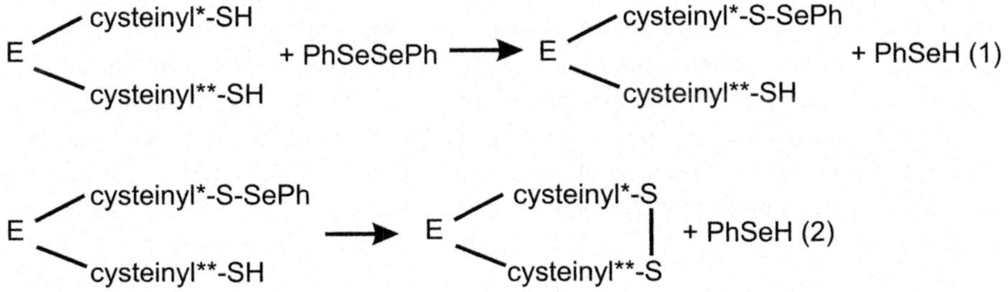

Fig. 4.8 Proposed molecular mechanism of oxidation of catalytic thiols from δ-ALAD by $(PhSe)_2$. Adapted from Farina et al. (2002)

4.5.1 An Overview of Organoarsenic (Group a Human Carcinogen)

Arsenic ions react with carbon and hydrogen to form organic arsenic. While organic arsenic compounds are mostly found in fish and shellfish, inorganic arsenic can be found in groundwater, which is used for drinking in many countries across the world. The Environmental Protection Agency (EPA) has categorized inorganic arsenic, which is present in soil and water, as a Group A human carcinogen (USEPA 2007). The toxicological effects of high concentrations of organic arsenic can be analogous to those of low concentrations of inorganic arsenic (ATSDR 2007). Methylated arsenic has a half-life of 30 hours, and ingested inorganic arsenic has a biological half-life of roughly 10 h, with 50.00–80.00% excreted in approximately 3 days (Gwaltney-Brant 2002). In seawater (2–3 μg dm^{-3}) and marine biota (up to 100 ng g^{-1} of wet weight), arsenic is primarily found as inorganic arsenate. It is also present in marine animals as arsenobetaine and in marine algae (Fig. 4.8) as dimethyl arsinyl ribosides, where arsenobetaine is absent. The main applications of arsenical derivatives are as high-purity metallic arsenic for the electronics industry, fining agents in the glass manufacturing process, metallic arsenic in nonferrous alloys, industrial chemicals (wood preservatives), and agricultural chemicals (herbicides and desiccants), and finally, a tiny amount of arsenic is used as an additive in animal food. Consumed arsenic has the potential to cross the placenta and cause concentrations in the cord blood that are similar to those in the mother's blood. After being absorbed, arsenites are converted to arsenates and methylated. Arsenic is carried throughout the body by blood, where it binds to red blood cells. As^{+3} is mainly enzymatically methylated in the liver to produce monomethylarsinic acid (MMA) and dimethylarsinic acid (DMA); dimethylated arsenic metabolites can then be produced by repeating this process. As (III), As (VI), MMA, and DMA are the main forms of arsenic that are quickly eliminated from the body through urine; DMA is typically the predominant form (ATSDR 2007; Fowler et al. 2022). Arsenic prefers sulfhydryl-rich keratin (skin) and tends to concentrate in the skin; however, chronic exposure is primarily responsible for deposits in the bones, teeth, hair, and nails (Gwaltney-Brant 2002). Urine, blood, hair, and nails are the most often utilized biomarkers for identifying and measuring arsenic exposure. While the primary indicators of acute arsenic toxicity are acute onset of profuse vomiting, diarrhea, colic, salivation, fever, and disruptions of the cardiovascular and central nervous systems, which may result in death, chronic exposure to inorganic arsenic compounds may cause neurotoxicity of both the peripheral and central nervous systems. In contrast, the primary symptoms of acute arsenic toxicity include fever, salivation, diarrhea, colic, vomiting, and disruptions of the central nervous system and cardiovascular system, which can be fatal (Gwaltney-Brant 2002). As a known carcinogen, arsenic is primarily linked to cancers of the skin, kidney, bladder, and lungs (ATSDR 2007).

Inorganic forms, the organic arsenic found in food and seafood, seem to be far less harmful (Fowler et al. 2022). Numerous fish species, including carp, mullet tuna, red snapper, bluefish, porgie, and salmon, have been found to contain arsenic. The Neretva River in Croatia has five fish species whose concentrations of arsenic have been measured. Organoarsenic intakes of roughly 0.050 mg kg^{-1} bw day^{-1} were established by the joint FAO/WHO Expert Committee on food additives (JECFA) as a provisional tolerable weekly intake (PTWI) that did not appear to be linked to any harmful effects. In many human populations, seafood, which includes shellfish, finfish, and seaweed, is the main source of arsenic (As) exposure. Foods derived from marine sources are mostly found in the form of organic compounds, as opposed to the dominant presence of inorganic as in water and many terrestrial foods (Fowler et al. 2022).

4.6 Human Health Risk Studies

4.6.1 Tolerable Average Residue Level (TARL)

TARLs have been suggested as the basis for governments to set the maximum residue limit (MRL) of TBT in seafood to ensure the health of the population (Belfroid et al. 2000). Penninks (1993), determined the tolerable daily intake (TDI) as 0.25 mg/kg BW/ day for TBT based on the immunotoxicity of rats. As well, the acceptable daily intake (ADI) of TPT compounds for humans has been set at 0.5 mg/kg BW/day. As TBT, DBT, and TPT have similar immune toxic effects, Guérin et al. (2007) suggested that it is reasonable to determine an ADI for the above organotins. Therefore, the European Food Safety Authority (EFSA 2004) established the TDI as 0.25 mg/kg BW/day based on the above assumption. The TARL, the tolerable value of organotin for the average consumer with an average weight of 60 kg, can be a useful tool to ensure the health of the population by comparing the TARL values and residue levels of TBT and TPT in seafood for each country (Belfroid et al. 2000). TARL was calculated as follows Eq. (4.25):

$$\text{TARL} = \sum \frac{\text{Tolerable daily intake} \times 65\,\text{kg BW}}{\text{average daily seafood consumption}} \tag{4.25}$$

For example, the TDI of TBT was 0.25 mg/kg BW/day and that of TPT was 0.5 mg/kg BW/day. The average daily seafood consumption is 0.067 kg/day in Taiwan. Therefore, TARLs were set at 243 ng/g wet weight for TBT and 485 ng/g wet weight for TPT. Thus, if people eat more seafood, their TARL would be less than those who consume smaller quantities.

4.6.2 Noncarcinogenic and Cancer Risk (CR) Assessment of Daily Intake for Organotin

The formula used to calculate the average daily dose ($ADD_{ingestion}$) was as follows (Lee et al. 2016):

$$ADD_{ingestion}\left(\mu g / kg / week\right) = \frac{C \times IR_{ingestion} \times AF \times 7\,days}{BW} \tag{4.26}$$

C TBT + DBT + TPT concentration (μg/g) in sample, $IR_{ingestion}$ daily consumption of rice (g/day) in each age group, AF absorption factor (%) (100% was used in the present study), *BW* body weight (kg).

4.6.3 Cancer Risk (CR)

It suggests the acute toxicity level of the pollutant and can be calculated as

$$CR = ADD \times SF \tag{4.27}$$

SF the cancer slope factor, differs depending on the involvement of organometal(loid) s (Majumdar et al. 2024).

4.6.4 Case Study (Health Risk Assessments of TBT and TPT)

Due to the adverse effects of OTs, the toxicities caused by TBT and TPT have been widely studied over the past few decades (Deng et al. 2018). The health risks associated with TBT and TPT in the effluents of the waterworks in the TGRR were assessed in this study: the Hazard Quotient (HQ) and Hazard Index (HI) approach. The following formula was used to calculate the HQ (USEPA 2007):

$$HQ = \Sigma_i ADD_i / RfD_i \tag{4.28}$$

More specifically, the hypothesis of dose additivity was calculated by summarizing the individual HQ values:

$$HI = \sum HQs = HQ1 + HQ2 + HQ3 + \cdots + HQn \tag{4.29}$$

where the RfD_i is the reference dose for TBT + DBT + TPT is 0.25 μg/kg BW/day (EFSA 2004) or TPT (0.5 μg/kg BW/day.

The following formulas (Eqs. (4.29)–(4.31)) were used to calculate the ADD*i*As the average daily dose (*i* represents one of the three routes: eating fish (F), drinking water (W), and skin absorption (S)) (μg (kg BW day)$^{-1}$), average daily intake of drinking water (ADDw), and average daily intake via the dermal route (ADDs), respectively (USEPA 2007):

$$\text{ADDF} = \text{C} \times \text{IR}_{\text{F}} \times \text{AF} / \text{BW} \tag{4.30}$$

$$\text{ADD}_w = \text{C} \times \text{IR}_w \times \text{ED} / (\text{BW} \times \text{AT}) \tag{4.31}$$

$$\text{ADD}_{\text{d}} = C \times k_p \times \text{ED} \times \text{SA} \times 24 \times 10^{-3} / (\text{BW} \times \text{AT}) \tag{4.32}$$

where IR_w is the ingestion rate of drinking water (2 L day^{-1}); ED is the exposure duration (30 a); BW is the body weight (60 kg, based on the average BW of adults in China); AT is the average exposure time (30 a); SA is the body surface area (16,600 cm^2); and k_p is the dermal permeability coefficient (0.001 cm h^{-1}) (USEPA 2007; EFSA 2004). Regarding the concentrations in the waterworks, the worst-case assessments for human exposure to TBT and TPT were determined. The LOD of 1.67 ng Sn L^{-1} was used for TBT because it was not detected; the highest concentration of TPT in the effluents of the waterworks was 4.14 ng Sn L^{-1}.

4.6.5 *Toxic Unit (TU)*

The TU estimates mixture-toxicity risks to both terrestrial and aquatic communities. We assessed the total exposure from the sediment by summing the exposures for all of the contaminants (Nolte et al. 2020):

$$\text{TU} = \sum_{i=1}^{n} \frac{[Ci]aq}{\gamma [EC50i]aq} \tag{4.33}$$

Ci is the concentration of organometallic compound and γi = 1 reflects all chemicals to have a similar mode of action (MoA) and negligible interaction between chemicals.

4.6.6 *Screening Level of OTC and Risk Quotient (Dai et al. 2022)*

Equations (4.34) and (4.35) were used to assess the risk posed to human health by OTCs (USEPA 2007). The noncarcinogenic risk posed by OTCs to human health can also be further divided into high, intermediate, and low risks according to the RQ II levels, RQ II > 1.0, 0.1 < RQ II < 1.0, and RQ II < 0.1, respectively:

$$\text{SL} = (\text{THQ} \times \text{AT} \times \text{BW} \times R\mathit{f}\text{Do}) / (\text{IRW} \times \text{EF} \times \text{ED}) \tag{4.34}$$

$$\mathrm{RQ\,II} = \sum_{i=1}^{n} RQi = \sum_{i=1}^{n} \frac{Ci}{SLi} \tag{4.35}$$

where SL is the OTC screening level (μg L^{-1}) and THQ is the target hazard quotient, with a default value of 1. Noncarcinogen averaging time (Atn, *d*) = ED [exposure duration (*y*)] × 365 day/year. Body weight (BW) and the tap water ingestion rate (IRW) were 65 kg and 1.7 L day^{-1}, respectively. EF (exposure frequency) was 350 days $year^{-1}$. RQ II is the risk quotient, and Ci is the concentration of each OTC. RfDo (mg·kg^{-1} BW day^{-1}) is the oral reference dose of OTCs, which is often replaced by the tolerable daily intake (TDI) due to the lack of RfDo values for OTCs. No TDI values for MMT, MBT, and TBT were found. The TDI values of TBT and TPhT were 0.25 μg kg^{-1} BW day^{-1} and 0.5 μg kg^{-1} BW day^{-1}, respectively (Penninks 1993). The TDI values of MBT, DBT, MPhT, and DPhT were 5.42 μg kg^{-1} BW day^{-1}, 1.75 μg kg^{-1} BW day^{-1}, 10.84 μg kg^{-1} BW day^{-1}, and 3.50 μg kg^{-1} BW day^{-1}, respectively, and were calculated using an in vitro immunotoxicity exposure test (Whalen et al. 2000).

4.6.7 Case Study (Annual Human Exposure to Methylmercury from Food in China)

Biotransport of MeHg through interregional food trade and other related processes is reported. Sunderland et al. (2018) analyzed sources of MeHg ingested by inhabitants among different age groups in Beijing based on the survey results regarding the production area of each food sample. The analysis was established based on the source-sink relationship of ingested MeHg transported from production to consumption (Liu et al. 2020):

$$\mathrm{Intake}_{i,k}\left(x,y\right) = \Sigma\left(C_{j,k}\left(X\right) \times I_{\mathrm{ij}}\left(y\right) \frac{A_{i,j}}{N_i} \times\right) \times P_i \times \frac{365}{10^{12}} \tag{4.36}$$

$\mathrm{Intake}_{i,k}$ (*x*, *y*) is the probabilistic distribution of the amount of MeHg (kg/year) ingested by group *i* (e.g., according to age and gender) contributed from source *k* (international, interprovincial, local, and unknown sources). $C_{j,k}$ (*x*) is the probabilistic distribution of the MeHg concentration (ng/g) of food j from source *k*; $A_{\mathrm{j,k}}$ is the amount of food *j* from source *k* based on the field survey; N_i is the total amount of food *j* collected in the study area; and P_i is the population number of group *i* (Liu et al. 2020).

4.7 Ecological Risk Assessment

Ecological models are now useful instruments for assessing both direct and indirect impacts as well as for calculating and controlling ecological hazards. Furthermore, ecological models may be used to assess hazards in situations where field trials are not possible, such as when a novel chemical is released into the environment, or to predict future potential dangers. When there is insufficient proof of harmful exposure, they might help reconstruct historical events or evaluate alternate hypotheses. When integrating pollutants into ecological models, the primary concern is often determining how probable it is that the harmful consequences seen at the individual level would affect the environment and populations.

Ecotoxicological effects depend on the bioavailable fraction of pollutants, and concentrations at the target sites induce molecular effects that propagate to various toxic manifestations in organisms (Escher et al. 1997; Fent 2003). Thereby, diverse aspects such as mechanisms of toxic action and ecological processes in contaminated.

systems are regarded in Fig. 4.9 (Fent 1998).

The risk quotient (RQ) was calculated to assess the ecological risk of TBT and TPhT for organisms in aquatic ecosystems (Leung et al. 2006):

$$RQ = PEC / PNEC \tag{4.37}$$

Here, PEC represents the exposure concentration (ng Sn L^{-1}) in the water phase predicted by the model. PNEC represents the predicted no-effect concentration (ng

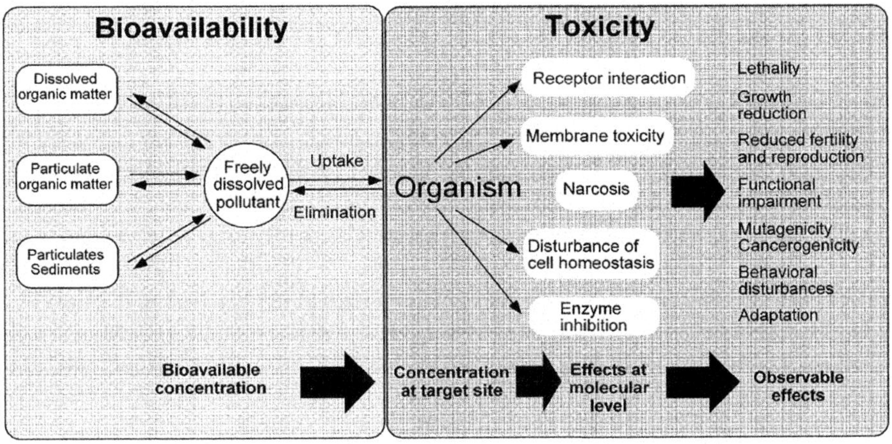

Fig. 4.9 (**a**) Ecotoxicological effects of pollutants at the target sites induce molecular effects that propagate to various toxic manifestations in organisms (Escher et al. 1997; Fent 2003). (**b**) Environmental risk assessment. PEC, predicted environmental concentration; PNEC, predicted no-observed-effect concentration. Risk analysis is necessary when PEC/PNEC>1 (Fent 2003)

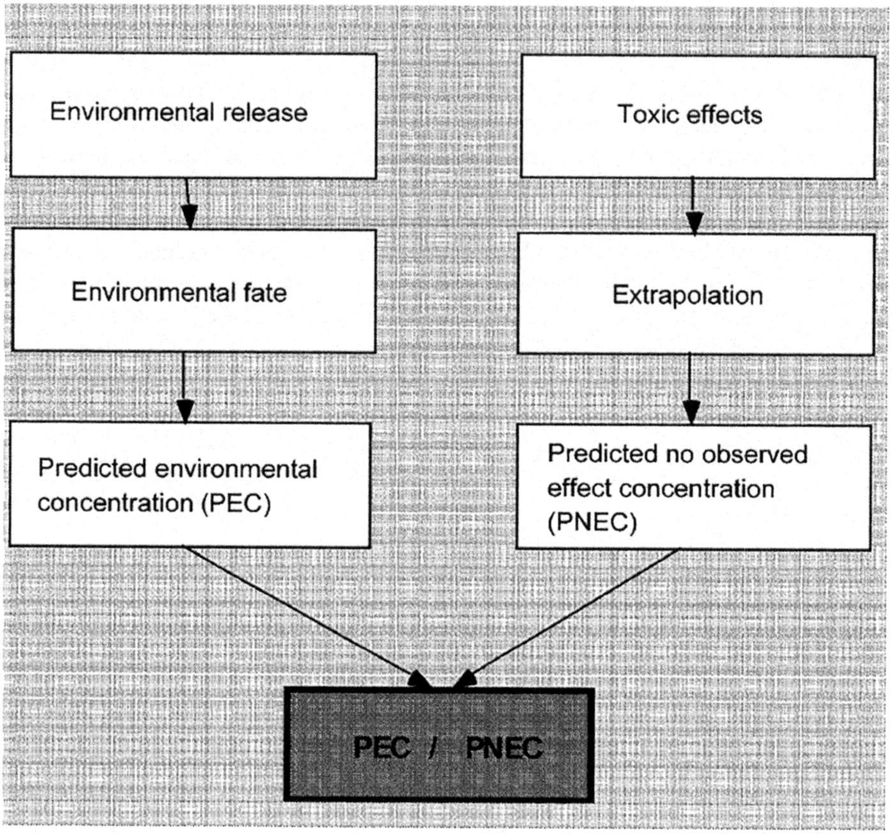

Fig. 4.9 (continued)

Sn L^{-1}), which is from water quality criteria (WQC) and toxicity thresholds in published literature. If RQ ≥ 1, the organisms in the aquatic environment are at significant risk. If 0.3 ≤ RQ < 1, there is a potential risk. If RQ < 0.3, there is no ecological risk (Leung et al. 2006). PEC/PNEC is a powerful integrative measure of the toxicity of chemicals not be achieved by analytical chemical measurements. The ecotoxicological topic requires an interdisciplinary approach considering physicochemical, molecular, toxicological, physiological, and ecological processes (Fig. 4.9a, b).

One of the main objectives of ecotoxicological studies is the prediction of chemical toxicity and the evaluation of the effects of chemicals on the environment. Since many modifying elements affect ecotoxicological consequences, comprehensive knowledge necessitates a methodical arrangement and progressive derivation of principles. Both the biophysicochemical characteristics of the chemical and the characteristics of the abiotic and biotic environments have an impact on the destiny and effects of a pollutant. Thus, it is necessary to create and integrate analytical instruments that are chemical, physical, and biological. The traditional method for calculating the risk associated with a pollutant involves estimating the exposure concentration and contrasting it with the no-effect level that is expected to be generated from both acute and long-term toxicity studies. There are several drawbacks to this empirical method, especially when it comes to substances with distinct mechanisms of action (van Leeuwen and Hermens 1996).

4.8 Biosensing-Bioindicator Potential of OMCs in Aquatic Systems

Biomarkers offer one important way in which environmental contamination effects can be monitored. Biomarkers are defined as the measurements of body fluids, cells, or tissues that indicate, in biochemical or cellular terms, the presence of contaminants or the magnitude of the host response (Bodin et al. 2004). According to Odendaal and Reinecke (2003), the "biomarkers" are any biological response to an environmental chemical that is measured inside an organism or its products (urine, feces, hairs, feathers, etc.) and indicates a departure from the normal status. A response may result from a biochemical, physiological, histological, and/or morphological (including appearance, pigmentation, and surface deformation) measurement of health, although behavioral effects are excluded. The utility of biomarkers derives from two features. First, organisms that inhabit specific ecological niches may biosynthesize unique biomarker compounds that can be used as indicators of the source organisms and their habitat. Second, organisms often possess an ability to regulate patterns of constituent biochemicals to remain viable under changing environmental conditions, so that environmental characteristics may be inferred from these patterns. To be a robust paleoenvironmental proxy, the signal carried by biomarkers must survive during deposition and burial in sediments over the time frame of interest.

OMCs have distinctive characteristics that render them exceptionally responsive to alterations in their surroundings, hence enabling their use as sensors for the detection of diverse chemicals or circumstances. In addition, the adaptability of organometallic compounds allows for their incorporation into various sensor platforms, rendering them appropriate for a diverse array of applications such as environmental surveillance. The significance of organometallic compounds in the fields of chemistry, electrochemistry, and biology can be attributed to their ability to function as sensors (Omer et al. 2024). Biomarkers have been developed to help monitor the effects of contaminants in marine invertebrates. Among these are the following biomarkers that have been used to assess the toxicity of TBT: metallothionein induction, acetylcholinesterase inhibition, imposex, lysosomal enlargement, lysosomal membrane destabilization, peroxisome proliferation, lysosomal activity, genetic or molecular biomarkers, TBT sensitive immunological biomarkers, apoptosis induction, phagocytic index, and amoebocytic index.

4.8.1 Biomarkers of TBT in Marine Invertebrates

4.8.1.1 Case Study 1: (Molecular Genetics)

Because pollutants interact with the receptors of organisms at the molecular level to cause their effects, the measurement of certain molecular biomarkers may have obvious advantages for detecting early chemical effects (Nicholson and Lam 2005). Schroth et al. (2005) utilized a strategy that identified molecular biomarkers and linked the study of abiotic stress to evolutionary history. These authors used the moon jellyfish, *Aurelia* spp., as a model species. The authors used complementary DNA subtraction analysis to identify genes that were differentially regulated after exposure to the chemical stressor TBT. They also identified differential expression patterns following exposure to TBT at different temperatures. Results suggested that the identified genes were involved in the response to the chemical.

4.8.1.2 Case Stdy 2: (Imposex)

Imposex is characterized by the development of morphological features. High incidences of imposex were characterized by lower female-to-male ratios, suggesting that sterility and female mortality were TBT related (Marshall and Rajkumar 2003). TBT generally shows the greatest accumulation among the butyltin compounds and is the primary cause of imposex (Sousa and Pastorinho 2017). Pessoa et al. (2001) studied the occurrence of organotin compounds in Portuguese coastal waters and found that acute effects from TBT were induced at concentrations as low as 1 μg/L in aquatic organisms; moreover, imposex was induced at levels below 0.5 ng/L of TBT. TBT at 20 ng/L caused sterility, and this was followed by the disappearance of the most sensitive neogastropods on a given shore. The use of imposex was the most sensitive indicator of exposure to TBT of all known nontarget pathological conditions. It became necessary to elucidate the effect mechanisms and the MoA involved once imposex was discovered in the early 1970s. Early research focused on potential inhibitory and modulatory interactions between TBT and detoxifying enzyme systems in various model species, including glutathione S-transferases and the cytochrome P-450-dependent monooxygenase system, for example (Fish et al. 1976). Eventually, the focus of the study shifted to endogenous hormone regulatory systems (Lee 1991) and then to nuclear and intracellular receptor-mediated control mechanisms of gene transcription (Evans 1988). Three main hypotheses about the imposex MoA have been identified throughout the subsequent years: the enhanced MoA of testosterone, the neuropeptide activation of APGWamide, and the agonism/activation of retinoid X-receptor (RXR) MoA. A quick synopsis of these three study topics is provided below. The "increased testosterone" or "increased androgen" hypothesis was the first significant attempt to explain the imposex MoA. It postulated that TBT worked by preventing the P-450-dependent aromatase system (CYP19A) from converting androgens to estrogens, which caused a shift in the point, the Monte Carlo method may also be utilized to evaluate the degree of uncertainty in the environmental elements as well as the model equations.

The assay simulations produce an effects matrix that modifies growth rates for populations in the food web model. The population effects from several simulations are used in calculating the risk (modified from Bartell 1996). One of the main objectives of ecotoxicological studies is to predict the toxicity of chemicals and evaluate their effects on the environment. For a comprehensive understanding, the nearly infinite number of modifying elements that cause ecotoxicological consequences needs a systematic arrangement and step-by-step deductive principles. Both the abiotic and biotic environment's characteristics and the chemical's biophysicochemical characteristics affect a pollutant's destiny and impact. Therefore, it is necessary to create and integrate analytical methods that are chemical, physical, and biological. Since concentration-effect relationships form the basis of ecotoxicological evaluations, it is clear how important concentration measurements and the selection of effects to be studied are. The availability of chemicals to organisms is influenced by several circumstances. Linking biological responses to the amount of a chemical that organisms absorb is one method to address these complications. When effective concentrations are stated as body residue or tissue-specific residues rather than in terms of the exposure concentration, it may be expected that observed disparities in the interspecific sensitivity to chemicals will be less prominent. In addition to whole-body concentration (physiologically based toxicokinetic modeling), the dynamics of the chemicals in various tissues and compartments should be considered for compounds that act specifically. The in vitro systems discussed here help simplify this method without sacrificing its applicability. To use the data from in vitro systems to simulate biological processes, further investigation is still needed. Estimating the exposure concentration and comparing it to the expected no-effect threshold estimated from acute and chronic toxicity studies is the traditional method of calculating the risk associated with a pollutant. There are several drawbacks to this empirical method, especially when it comes to compounds with distinct mechanisms of action (van Leeuwen and Hermens 1996). Mode-of-action end points should be added to lethality and other physiological end points for this class of drugs.

4.10 Historical Overview of the Ecotoxicity Concept

For almost 90 years, particularly the GRs—RMgHal—have been essential to synthesizing organomagnesium compounds. Their capacity to add to polar groups like C=O and -C≡N is especially useful. Because beryllium oxide, produced when organoberyllium compounds burn, is poisonous when inhaled as dust, they have not yet found any substantial synthetic uses. Compared to the derivatives of magnesium and lithium, they are often less reactive, less stable, and less accessible. Compared to GRs, their behavior is less complicated by side reactions. The organic chemistry of barium, strontium, and calcium is even less understood. On the other hand, compounds of organozinc, cadmium, and mercury have been recognized for more than a century and are utilized extensively in synthesis. In Germany and Japan, acetaldehyde was previously produced by hydrating ethyne. At around 70 °C, ethylene was converted to 20% aqueous sulfuric acid while iron and mercury salts were present. The mechanism might have anything to do with the responses covered above (Albers and Coville 1984).

4.10.1 Minamata Disease

There were severe repercussions, including Minamata, Kyushu, Japan, when the sea was exposed to Hg^{2+} from this procedure. It was changed to produce extremely hazardous methylmercury compounds anaerobically, which were absorbed by crustacea and fish as MeHgSMe. Several people lost their lives after consuming the fish. $MeHg^{+}$ is a very mild Lewis acid that forms a strong combination with sulfur and often has two coordinates, like the 1-cysteine complex ($MeHgSCH_2CH(NH_3^{+})CO_2^{-}\cdot H_2O$), and has a great affinity for nucleosides and nucleotides and polynucleosides as well. Organomercury compounds have been utilized as diuretics, general antiseptics, and antisyphilitic medications. Although they are still utilized in other applications, they have been supplanted in the first place by contemporary antibiotics. Phenylmercury acetate is among the most basic organomercury antiseptics (Powell 1988).

According to some ecotoxicologists, TBT is the most dangerous man-made chemical ever purposefully put into the environment in significant amounts. Over more than 50 years of research, thousands of studies about TBT's environmental fate and effects have been reported. This field continues to garner significant attention, largely because of TBT's actions as an endocrine-disrupting compound that can potentially affect biological fitness (such as growth, development, and reproduction) in sensitive species groups. In this sense, the emphasis will be on determining the molecular processes and the modes of action (MoA) that TBT has on susceptible species, as well as how these mechanisms could be pertinent to the development of toxicity in other marine species (Beyer et al. 2022). It became necessary to elucidate the effect mechanisms and the MoA involved once imposex was discovered in the early 1970s. Early research focused on potential inhibitory and modulative interactions between TBT and detoxifying enzyme systems in various model species, including glutathione S-transferases and the cytochrome P-450-dependent monooxygenase system, for example (Fish et al. 1976). Eventually, the focus of the study shifted to endogenous hormone regulatory systems (Lee 1991) and then to nuclear and intracellular receptor-mediated control mechanisms of gene transcription (Evans 1988).

4.11 Cutoff Point Criteria for Indexing OMC Pollution

Direct results of the pollutant release into aquatic environments differ depending on the degree and length of exposure and are often investigated to determine the danger associated with a particular chemical and the determination of a cutoff point for its allowable concentrations. These investigations are often founded on toxicity testing conducted in laboratories utilizing model species' reactions to exposure to a single pollutant. Usually, direct impacts cause an organism's abundance to decrease, for instance, a rise in mortality, a decline in fertility, or a growth rate, etc. Pollutants, however, can still affect tolerant species through other ecological mechanisms. For example, direct impacts of contaminants on predators can change competitive interactions, which in turn modify the abundance and dynamic behavior of resistant species in other trophic levels. According to Fleeger et al. (2003), these impacts are known as indirect (or secondary) contaminant effects, and occasionally, they might have an even greater impact than the direct (toxic) consequences of a contaminant. According to Bartell (1996) and Pastorok et al. (2003), ecological models are now useful instruments for assessing both direct and indirect impacts as well as for calculating and controlling ecological hazards. Furthermore, ecological models may be used to assess hazards in situations where field trials are not possible, such as when a novel chemical is released into the environment, or to predict future potential dangers. When there is insufficient proof of harmful exposure, they might help reconstruct historical events or evaluate alternate hypotheses. When integrating pollutants into ecological models, the primary concern is often determining how probable it is that the harmful consequences seen at the individual level would affect the environment and populations. In risk assessment, the ecological impact of toxicant stress has traditionally been measured by an organism's mortality (survival rate) (Bartell 1996); however, more recent methods have emerged. Tanaka (2003), for instance, suggested using population vulnerability analysis (PVA), which entails calculating the likelihood of extinction by assessing the detrimental effects of polluting substances on the intrinsic rate of natural growth (Ginzburg et al. 1982).

The overall strategy and associated decision logic for the evaluation of organometallic (OM) and organometallic salt (OMS) compounds are illustrated in Fig. 4.12. Further elaboration and detail on considerations associated with each of the steps, as well as how to deal with different levels of data availability, are studied by OECD (2017).

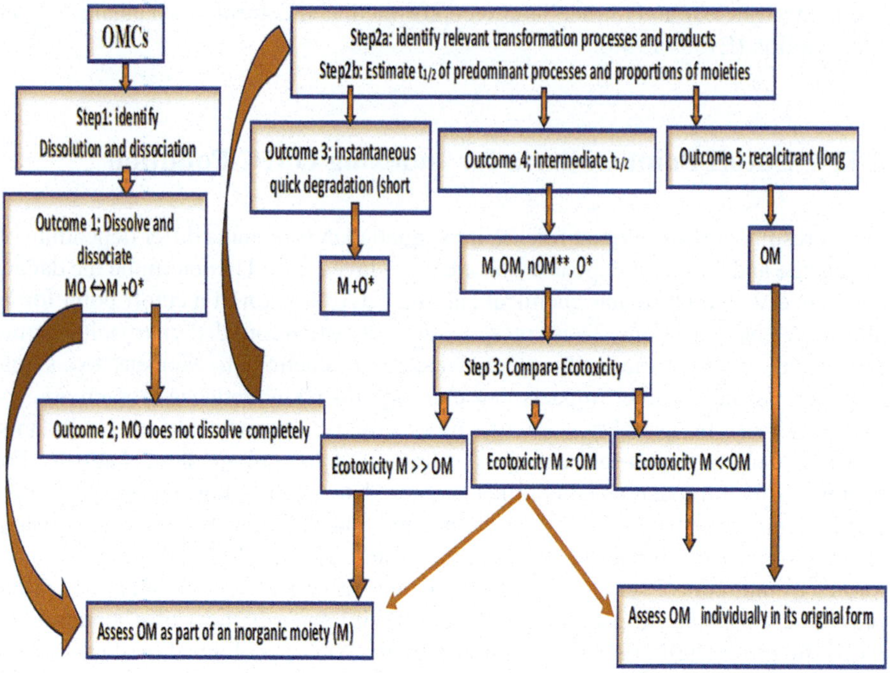

Fig. 4.12 Cutoff points of the ecological assessment strategy for OMCs (USEPA 2007; OECD 2017). *OMS* organic metal salt, *OM* organometallic, *M* metal/metalloid (e.g., cation, oxyanion, free ion) or metal/metalloid free inorganic moiety, *O** organic moiety, *nOM*** transformation product that is a new organometallic, *t1/2* half-life; **Any O* released as a product upon dissociation or another transformation process should be assessed as part of an organic assessment (e.g., matched to an organic moiety or group assessment or considered immediately if the OM/OMS is being addressed on a substance-by-substance basis) if warranted (option not illustrated in the flowchart), ***Any nOM* released as a product upon the transformation of the original OM or OMS should reenter Step 2

References

Abadin HG, Pohl HR (2010) Alkyl lead compounds and their environmental toxicology. Met Ions Life Sci 7:153–164. https://doi.org/10.1039/BK9781847551771-00153

Albers MO, Coville NJ (1984) Reagent and catalyst induced substitution reactions of metal carbonyls. Coord Chem Revs 53:227

Ali MM, Ali ML, Rakib MR, Islam MS, Bhuyan MS, Senapathi V, Chung SY, Roy PD, Sekar S, Towfiqul Islam AR, Rahman MZ (2022) Seasonal behavior and accumulation of some toxic metals in commercial fishes from Kirtankhola tidal river of Bangladesh–a health risk taxation. Chemosphere 301:134660

Antizar-Ladislao B (2008) Environmental levels, toxicity and human exposure to tributyltin (TBT)-contaminated marine environment. A review. Environ Int 34:292–308

ATSDR (2007) Toxicological profile for arsenic. Agency for Toxic Substances and Disease Registry. U.S. Department of Health and Human Services, Atlanta. Available http://www.atsdr.cdc.gov/ToxProfiles/tp2.pdf

Aziz KN, Mahmood A et al (2024) Organometallic complexes and reaction methods for synthesis: a review. Rev Inorg Chem. https://doi.org/10.1515/revic-2024-0037
Bangkedphol S, Keenan H, Davidson C, Sakultantimetha A, Songsasen A (2009) The partition behavior of tributyltin and prediction of environmental fate, persistence and toxicity in aquatic environments. Chemosphere 77:1326–1332
Bartell SM (1990) Ecosystem context for estimating stress-induced reductions in fish populations. Am Fish Soc Symp 8:167–182
Bartell SM (1996) Ecological and environmental risk assessment: theory and practice. In: Kolluru R, Bartell SM, Pitblado R, Stricoff S (eds) Risk assessment and management handbook. McGraw-Hill, New York, p 1050
Batista-Andrade JA, Caldas SS, Batista RM (2018) From TBT to booster biocides: levels and impacts of antifouling along coastal areas of Panama. Environ Pollut 234:243–252
Belfroid AC, Purperhart M, Ariese F (2000) Organotin levels in seafood. Mar Pollut Bull 40:226–232
Bettin C, Oehlmann J, Stroben E (1996) TBT-induced imposex in marine neogastropods is mediated by an increasing androgen level. Helgol Meeresunters 50:299–317
Beyer J, Song Y, Tollefsen KE, Berge JA, Tveiten L, Helland A, Øxnevad S, Schøyen M (2022) The ecotoxicology of marine tributyltin (TBT) hotspots: a review. Mar Environ Res 179:105689. https://doi.org/10.1016/j.marenvres
Bodin N, Burgeot T, Stanisiere JY, Bolquene G, Menard D, Minier C, Boutet I, Amat A, Chard Y, Budzinski H (2004) Seasonal variations of a battery of biomarkers and physiological indices for the mussel Mytilus galloprovincialis transplanted into the Northwest Mediterranean Sea. Comp Biochem Physiol Toxicol Pharmacol 138(4):411–427
Borgert CJ, Quill TF, Carty LS, Mason AM (2004) Can mode of action predict mixture toxicity for risk assessment? Toxicol Appl Pharmacol 201:85–96
Bouchard N, Pelletier E, Fournier M (1999) Effects of butyltin compounds on phagocytic activity of hemocytes from three marine bivalves. Toxicol Chem 18(3):519–522
Boyer PD (1997) The ATP synthase—a splendid molecular machine. Annu Rev Biochem 66:717–749
Burton ED, Phillips IR, Hawker DW (2004) Sorption and desorption behavior of tributyltin with natural sediments. Environ Sci Technol 38:6694–6700
Cao D, Jiang G, Zhou Q, Yang R (2009) Organotin pollution in China: an overview of the current state and potential health risk. J Environ Manage 90(S1):7. https://doi.org/10.1016/j.jenvman.2008.06.007
Carson P (2002) Hazardous chemicals handbook.2nd edn. Butterworth-Heinemann, Oxford
Castro LFC, Lima D, Machado A, Melo C, Hiromori Y, Nishikawa J, Nakanishi T, Reis-Henriques MA, Santos MM (2007) Imposex induction is mediated through the retinoid X receptor signalling pathway in the neogastropod Nucella lapillus. Aquat Toxicol 85:57–66
Çetintürk K, Ünlü S (2022) The first observation of antifouling organotin compounds and booster biocides in sediments from Samsun port area, Black Sea, Turkey. Mar Pollut Bull 176:113408. https://doi.org/10.1016/j.marpolbul.2022.113408
Champ MA, Seligman PF (1996) Organotin—environmental fate and effects. Chapman and Hall, London, pp 1–25
Chicano JJ, Ortiz A, Teruel JA, Aranda FJ (2002) Organotin compounds promote the formation of non-lamellar phases in phosphatidylethanolamine membranes. Biochim Biophys Acta 1558:70–81
Chow SC, Kass GE, McCabe Jr MJ, Orrenius S (1992) Tributyltin increases cytosolic free Ca^{2+} concentration in thymocytes by mobilizing intracellular Ca^{2+}, activating a Ca^{2+} entry pathway, and inhibiting Ca^{2+} efflux. Arch Biochem Biophys 298(1):143–149. https://doi.org/10.1016/0003-9861(92)90105-6
Cima F, Dominici D, Mammi S, Ballarin L (2002) Appl Organometal Chem 182:247

Cornelis R (2007) Handbook on the toxicology of metals. General chemistry, sampling, analytical methods, and speciation. Wiley, Hoboken, pp 11–38. https://doi.org/10.1016/b978-012369413-3/50057-4

Craig PJ (2023) Organometallic compounds in the environment.2nd edn. Wiley, Hoboken, p 434

Craig PJ, Jenkins RO (2004) Organometallic compounds in the environment: an overview. In: Hirner AV, Emons H (eds) Organic metal and metalloid species in the environment. Springer, Berlin. https://doi.org/10.1007/978-3-662-09135-7_1

Dai Q, Chen L, Li P, Xia S, Wang Y and Huang Q (2022), Occurrence and risk assessment of organotin compounds in the surface water of the upper Yangtze River Estuary. Front. Environ. Sci. 10:964486. https://doi.org/10.3389/fenvs.2022.964486

de Oliveira CR, dos Santos DM, Madureira LAD, de Marchi MRR (2010) Speciation of butyltin derivatives in surface sediments of three southern Brazilian harbors. J Hazard Mater 181(1-3):851–856. https://doi.org/10.1016/j.jhazmat.2010.05.091

Deng T, Wu L, Gao J-M, Zhou B, Zhang Y-L, Wu W-N, Tang Z-H, Jiang W-C, Huang W-L (2018) Occurrence and health risk assessment of organotins in waterworks and the source water of the three gorges reservoir region, China. Environ Sci Pollut Res 25(15):15019–15028. https://doi.org/10.1007/s11356-018-1704-0

Dom N, Penninck M, Knapen D, Blust R (2012) Discrepancies in the acute versus chronic toxicity of compounds with a designated narcotic mechanism. Chemosphere 87(7):742–749. https://doi.org/10.1016/j.chemosphere.2011.12

Dubey SK, Roy U (2003) Biodegradation of tributyltins (organotins) by marine bacteria. Appl Organomet Chem 17:3–8. https://doi.org/10.1002/aoc.394

EFSA (2004) Opinion of the scientific panel on contaminants in the food chain [CONTAM] to assess the health risks to consumers associated with exposure to organotins in foodstuffs. EFSA J 2(10):102. https://doi.org/10.2903/j.efsa.2004.102

Egorochkin AN, Kuznetsova OV, Khamaletdinova NM, Domratcheva-Lvova LG (2013) Toxicity of organometallic compounds: correlation analysis via substituent constants. J Organomet Chem 735:88–92. https://doi.org/10.1016/j.jorganchem.2013.03.029

Egorova KS, Ananikov VP (2017) Toxicity of metal compounds: knowledge and myths. Organometallics 36(21):4071–4090

Elias AJ, Gupta BD (2010) Basic organometallic chemistry. University Press (India) Pvt, Ltd, Hyderabad

Escher BI, Behrab R, Rik IL, Eggen C, Fent K (1997) Molecular mechanisms in ecotoxicology: an interplay between environmental chemistry and biology. Chimia 51:915–921

Evans RM (1988) The steroid and thyroid hormone receptor superfamily. Science 240:889–895

Fang LP, Borggaard OK, Marcussen H, Holm PE, Hansen HCB (2010) The pH-dependent adsorption of tributyltin to charcoals and soot. Environ Pollut 158:3642–3649

Fang L, Xu C, Li J, Borggaard OK, Wang D (2017) The importance of environmental factors and matrices in the adsorption, desorption, and toxicity of butyltins: a review. Environ Sci Pollut Res 24(10):9159–9173. https://doi.org/10.1007/s11356-017-8449-z

Farina M, Barbosa NBV, Nogueira CW, Folmer V, Zeni G, Andrade LH, Braga AL, Rocha JBT (2002) Reaction of diphenyl diselenide with hydrogen peroxide and inhibition of delta-aminolevulinate dehydratase from rat liver and cucum-485 ber leaves. Braz J Med Biol Res 35:623–631

Feng X, Zhang K, Hempenius MA, Vancso GJ (2015) Organometallic polymers for electrode decoration in sensing applications. RSC Adv 5(129):106355–106376. https://doi.org/10.1039/C5RA21256A

Fent K (1996a) Organotin compounds in municipal wastewater and sewage sludge: contamination, fate in the treatment process and ecotoxicological consequences. CRC Rev Toxicol 26:151–159

Fent K (1996b) Ecotoxicology of organotin compounds. Crit Rev Toxicol 26(1):3–117. https://doi.org/10.3109/10408449609089891

Fent K (1998) Effects of organotin compounds in fish: from the molecular to the population level. In: Braunbeck T, Hinton DE, Streit B (eds) Fish ecotoxicology, vol 86. Birkhäuser, Basel. https://doi.org/10.1007/978-3-0348-8853-0_8
Fent K (2003) Ecotoxicological problems associated with contaminated sites. Toxicol Lett 11:140–141. https://doi.org/10.1016/s0378-4274(03)00032-8
Fent K (2006) Worldwide occurrence of organotins from antifouling paints and effects in the aquatic environment. In: Konstantinou IK (ed) Antifouling paint biocides. Earth and environmental science: the handbook of environmental chemistry, vol 50. Springer, Berlin, pp 71–100. https://doi.org/10.1007/698_5_050
Fent K, Stegeman JJ (1991) Effects of tributyltin chloride in vitro on the hepatic microsomal monooxygenase system in the fish Stenotomus chrysops. Aquat Toxicol 20:159–168
Fish RH, Kimmel EC, Casida JE (1976) Bio-organotin chemistry - reactions of tributyltin derivatives with a cytochrome-P-450 dependent monooxygenase enzyme-system. J Organomet Chem 118:41–54
Fleeger JW, Carman K, Nisbet R (2003) Indirect effects of contaminants in aquatic ecosystems. Sci Total Environ 317: 207–233.
Fodor I, Urban P, Scott AP, Pirger Z (2020) A critical evaluation of some of the recent so-called 'evidence' for the involvement of vertebrate-type sex steroids in the reproduction of mollusks. Mol Cell Endocrinol 516:10
Fowler BA, Chou C-HSJ, Jones RL, Costa M, Chen C-J (2022) Arsenic. In: Nordberg BA, Costa M (eds) Handbook on the toxicology of metals, vol 2. 5th edn. Academic, New York, pp 41–89
Gadd GM (2000) Microbial interactions with tributyltin compounds: detoxification, accumulation, and environmental fate. Sci Total Environ 258(1-2):119–127. https://doi.org/10.1016/S0048-9697(00)00512-X
Ginzburg LR, Slobodkin LB, Johnson K, Bindman AG (1982) Quasiextintion probabilities as a measure of impact on populations growth. Risk Anal 2:171–181
Gomez NA, Sturla LJ, Ferrando A, Garrido M, Domini CE (2024) Update on the status of the contamination by organotin compounds in sediment of Nuevo gulf, Argentina. Insights from field and experimental studies. Mar Pollut Bull 200:116087. https://doi.org/10.1016/j.marpolbul.2024.116087
Guérin T, Sirot V, Volatier JL, Leblanc JC (2007) Organotin levels in seafood and its implications for health risk in high-seafood consumers. Sci Total Environ 388:66–77
Gwaltney-Brant M (2002) Heavy metals. In: Haschek WM, Rousseaux CG, Wallig MA (eds) Handbook of toxicologic pathology2nd edn. Academic, New York, pp 701–733. https://doi.org/10.1016/B978-012330215-1/50027-2
Haggera JA, Depledge MH, Galloway TS (2005) Toxicity of tributyltin in the marine mollusk Mytilus edulis. Mar Pollut Bull 51:811–816
Hamer K, Karius V (2005) Tributyltin release from harbour sediments–modelling the influence of sedimentation, bio-irrigation and diffusion using data from Bremerhaven. Mar Pollut Bull 50(9):980–992. https://doi.org/10.1016/j.marpolbul.04.007
Hoch M, Schwesig D (2004) Parameters controlling the partitioning of tributyltin (TBT) in aquatic systems. Appl Geochem 19(3):334. https://doi.org/10.1016/s0883-2927(03)00131-8
Horvat M, Nolde N, Fajon V, Jereb V, Logar M, Lojen S, Jacimovic R, Falnoga I, Liya Q, Faganeli J, Drobne D (2003) Total mercury, methylmercury and selenium in mercury polluted areas in the province Guizhou, China. Sci Total Environ 304(1-3):231–256. https://doi.org/10.1016/S0048-9697(02)00572-7
Hosmane NS (2017) Advanced inorganic chemistry. Organomet Chem 10:199–208. https://doi.org/10.1016/B978-0-12-801982-5.00010-2
House JE, House KA (2016) Descriptive inorganic chemistry. In: Organometallic compounds. Elsevier, Amsterdam, pp 371–393. https://doi.org/10.1016/B978-0-12-804697-5.00022-1
Huang JH, Matzner E (2004) Adsorption and desorption of organotin compounds in organic and mineral soils. Eur J Soil Sci 55:693–698

Huang W, Wu Q, Xu F, Li L, Li L, Que H, Zhang G (2020) Functional characterization of retinoid X receptor with an emphasis on the mediation of organotin poisoning in the Pacific oyster (Crassostrea gigas). Gene 753:144780

IPCS (1990) Environmental health criteria for tributyltin compounds, environmental health criteria No. 116. World Health Organization, Geneva, p 277

Jenkins RO (2007) Comprehensive organometallic chemistry III. In: Environmental and biological aspects of organometallic compounds. Elsevier, Amsterdam, pp 603–661. https://doi.org/10.1016/B0-08-045047-4/00178-3

Kucharski D, Stasiuk R, Drzewicz P, Skowronek A, Strzelecka A, Mianowicz K, Giebułtowicz J (2022) Fit-for-purpose assessment of QuEChERS LC-MS/MS methods for environmental monitoring of organotin compounds in bottom sediments of the Odra River estuary. Molecules 27:4847. https://doi.org/10.3390/molecules27154847

LaFont R (2000) Endocrinol Invertebr Ecotoxicol 9:41–57

LaLone CA, Villeneuve DL, Lyons D, Helgen HW, Robinson SL, Swintek JA, Saari TW, Ankley GT (2016) Sequence alignment to predict across species susceptibility (SeqAPASS): a web-based tool for addressing the challenges of cross-species extrapolation of chemical toxicity. Toxicol Sci 153:228–245

Landmeyer JE, Tanner TL, Watt BE (2004) Biotransformation of tributyltin to tin in freshwater river-bed sediments contaminated by an organotin release. Environ Sci Technol 38:4106–4112. https://doi.org/10.1021/es030697z

Laughlin Jr RB, Johannesen RB, French W, Guard H, Brinckman FE (1985) Structure-activity relationships for organotin compounds. Environ Toxicol Chem 4:343–351

LeBlanc GA, Campbell PM, den Besten P, Brown RP, Chang RP, Coats JR, deFur PL, Dhadialla T, Edwards J, Riddiford LM, Simpson MG, Snell TW, Thorndyke M, Matsumura F (1999) The endocrinology of invertebrates. In: Endocrine disruption in invertebrates: endocrinology, testing and assessment. SETAC, Brussels, pp 23–106

Lee RF (1991) Metabolism of tributyltin by marine animals and possible linkages to effects mar. Environ Res 32:29–35

Lee C-C, Hsu Y-C, Kao Y-T, Chen H-L (2016) Health risk assessment of the intake of butyltin and phenyltin compounds from fish and seafood in Taiwanese population. Chemosphere 164:568–575. https://doi.org/10.1016/j.chemosphere.2016.08

Lesoway MP, Henry JQ (2021) Retinoids promote penis development in sequentially hermaphroditic snails. Dev Biol 478:122–132

Leung KMY, Kwong RPY, Ng WC, Horiguchi T, Qiu JW, Yang R et al (2006) Ecological risk assessments of endocrine disrupting organotin compounds using marine neogastropods in Hong Kong. Chemosphere 65:922–938. https://doi.org/10.1016/j.chemosphere.2006.03.048

Liu M, Cheng M, Zhang Q, Hansen G, He Y, Yu C, Lin H, Zhang H, Wang X (2020) Significant elevation of human methylmercury exposure induced by the food trade in Beijing, a developing megacity. Environ Int 135:105392. https://doi.org/10.1016/j.envint.2019

Ma H, Dai S, Huang G (2000) Distribution of tributyltin chloride in laboratory simulated estuarine microcosms. Water Res 34:2829–2841

Majumdar A, Upadhyay MK, Ojha M, Biswas R, Dey S, Sarkar S, Moulick D et al (2024) A critical review on the organo-metal(loid)s pollution in the environment: distribution, remediation and risk assessment. Sci Total Environ 951:175531. https://doi.org/10.1016/j.scitotenv.2024.175531

Mandal A, Giri S, Giri A (2024) Assessment of toxicity, genotoxicity and oxidative stress in Fejervarya limnocharis exposed to tributyltin. Environ Sci Pollut Res 31:14938–14948. https://doi.org/10.1007/s11356-024-32220-5

Marshall DJ, Rajkumar A (2003) Imposex in the indigenous *Nassarius kraussianus* (Mollusca: *Neogastropoda*) from south African harbours. Mar Pollut Bull 46:1150–1155

Meador JP (1997) Comparative toxicokinetics of tributyltin in fve marine species and its utility in predicting bioaccumulation and acute toxicity. Aquat Toxicol 37:307–326. https://doi.org/10.1016/S0166-445X(96)00827-2

Meador JP (2000) Predicting the fate and effects of tributyltin in marine systems. Rev Environ Contam Toxicol 166:1–48
Meador JP, Krone CA, Dyer DW, Varanasi U (1997) Toxicity of sediment-associated tributyltin to in faunal invertebrates: species comparison and the role of organic carbon. Mar Environ Res 43:219
Mizuta T, Kubokawa K (2007) Presence of sex steroids and cytochrome P450 genes in amphioxus. Endocrinology 148:3554–3565
Nesci S, Ventrella V, Trombetti F, Pirini M, Borgatti AR, Pagliarani A (2011) Tributyltin (TBT) and dibutyltin (DBT) differently inhibit the mitochondrial Mg-ATPase activity in mussel digestive gland. Toxicol In Vitro 25(1):117–124. https://doi.org/10.1016/j.tiv.2010.10.001
Nicholson S, Lam PKS (2005) Pollution monitoring in Southeast Asia using biomarkers in the mytilid mussel Perna viridis (Mytilidae: Bivalvia). Environ Int 31:121–132
Nolte TM, De Cooman W, Vink JPM, Elst R, Ryken E, Ragas AMJ, Hendriks AJ (2020) Bioconcentration of organotin cations during molting inhibits Heterocypris incongruens growth. Environ Sci Technol 54(22):14288–14301. https://doi.org/10.1021/acs.est.0c02855
O'Neil RV, Gardner RH, Barnthouse LV, Suter GW, Hildebrand SG, Gehrs CW (1982) Ecosystem risk analysis: a new methodology. Environ Toxicity Chem 1:167–177
Odendaal JP, Reinecke AJ (2003) Quantifying histopathological alterations in the hepatopancreas of the woodlouse Porcellio laevis (isopoda) as a biomarker of cadmium exposure. Ecotoxicol Environ Saf 56(2):319–325. https://doi.org/10.1016/S0147-6513(02)00163-X
OECD (2017) Guidance on selecting a strategy for assessing the ecological risk of organometallic and organic metal salt substances based on their environmental fate. In: OECD series on testing and assessment, no. 212. OECD Publishing, Paris. https://doi.org/10.1787/9789264274785-en
Okoro HK, Fatoki OS, Adekola FA, Ximba BJ, Snyman RG, Opeolu B (2011) Human exposure, biomarkers, and fate of organotins in the environment. In: Whitacre D (ed) Reviews of environmental contamination and toxicology. Reviews of environmental contamination and toxicology, vol 213. Springer, New York. https://doi.org/10.1007/978-1-4419-9860-6_2
Omer P, Abdulkareem EI, Omer RA, Faruq RR (2024) A review of organometallic compounds as versatile sensors in environmental, medical, and industrial applications. Rev Inorg Chem. https://doi.org/10.1515/revic-2024-0055
Orrenius S, McCabe MJ, Nicotera P (1992) Ca^{2+}-dependent mechanisms of cytotoxicity and programmed cell death. Toxicol Lett 64:357–364. https://doi.org/10.1016/0378-4274(92)90208-2
Otto S (1996) Die Bedeutung von DOC fur den Kohlenstofffluss im Ozean. Berichte des Fachbereichs Geowissenschaften 87, Univ. Bremen, p 150
Panga MJ, Zhao Y (2024) Male reproductive toxicity of antifouling chemicals: insights into oxidative stress-induced infertility and molecular mechanisms of zinc pyrithione [ZPT]. Antioxidants 13:173. https://doi.org/10.3390/antiox13020173
Pascoal S, Carvalho G, Vasieva O, Hughes R, Cossins A, Fang YX, Ashelford K, Olohan L, Barroso C, Mendo S, Creer S (2013) Transcriptomics and in vivo tests reveal novel mechanisms underlying endocrine disruption in an ecological sentinel, Nucella lapillus. Mol Ecol 22:1589–1608
Pastorok RA, Ak¸cakaya R, Regan H, Ferson S and Bartell SM (2003). Role of ecological modeling in risk assessment, Hum. Ecol. Risk Assess 9939–972.
Penninks AH (1993) The evaluation of data-derived safety factors for bis(trinbutyltin) oxide. Food Addit Contam 10:351–361
Pessoa MF, Fernando A, Oliveira JS (2001) Use of imposex (pseudohermaphroditism) as indicator of the occurrence of organotin compounds in Portuguese coastal waters—Sado and mira estuaries. Environ Toxicol 16(3):234–241
Powell P (1988) Principles of organometallic chemistry.2nd edn. Chapman and Hall, London. https://doi.org/10.1007/978-94-009-1197-0
Powers MF, Beavis AD (1991) Triorganotins inhibit the mitochondrial inner membrane anion channel. J Biol Chem 266:17250–17256
Schaller CP, Grahm KJ, Johnson BJ (2015) Modules for introducing organometallic reactions: a bridge between organic and inorganic chemistry. J Chem Educ 92:986–992

Schroth W, Ender A, Schierwater B. Molecular biomarkers and adaptation to environmental stress in moon jelly (Aurelia spp.). Mar Biotechnol (NY). 2005 Sep-Oct;7(5):449-61. https://doi.org/10.1007/s10126-004-4095-9. Epub 2005 Jun 4. PMID: 15976937.

Scott AP (2012) Do mollusks use vertebrate sex steroids as reproductive hormones? Part I: critical appraisal of the evidence for the presence, biosynthesis and uptake of steroids. Steroids 77:1450–1468

Scott AP (2013) Do mollusks use vertebrate sex steroids as reproductive hormones? II. Critical review of the evidence that steroids have biological effects. Steroids 78:268–281

Smit AB, Jimenez CR, Dirks RW, Croll RP, Geraerts WPM (1992) Characterization of a cDNA clone encoding multiple copies of the neuropeptide APGWamide in the mollusk Lymnaea-stagnalis. J Neurosci 12:1709–1715

Sousa ACA, Pastorinho MR (2017) Organotins and imposex in Europe: a pre-ban and post-ban perspective. In: Horiguchi T (ed) Biological effects by organotins. Springer, Tokyo. https://doi.org/10.1007/978-4-431-56451-5_5

Sousa A, Genio L, Mendo S, Barrosoi C (2005) Comparison of the acute toxicity of tributyltin and copper to veliger larvae off Nassarius reticulatus (L.). Appl Organomet Chem 19:324–328

Sternberg RM, Gooding MP, Hotchkiss AK, LeBlanc GA (2010) Environmental-endocrine control of reproductive maturation in gastropods: implications for the mechanism of tributyltin-induced imposex in prosobranchs. Ecotoxicology 19:4–23

Sunday AO, Alafara BA, Oladele GO (2012) Toxicity and speciation analysis of organotin compounds. Chem Spec Bioavail 24(4):216–226. https://doi.org/10.3184/095422912X13491962881734

Sunderland EM, Li M, Bullard K (2018) Decadal changes in the edible supply of seafood and methylmercury exposure in the United States. Environ Health Perspect 126(1):017006

Tanaka Y (2003) Ecological risk assessment of pollutant chemicals: extinction risk based on population-level effects. Chemosphere 53:421–425

Temerdashev Z, Abakumov P, Bolshov M, Abakumova D, Pupyshev A (2023) Possibilities and limitations of ICP-spectrometric determination of the total content of tin and its inorganic and organic speciations in waters with different salinity levels—part 2: separate determination of inorganic and organic speciations of tin. Molecules 28:6615. https://doi.org/10.3390/molecules28186615

TOXNET (2017) Toxicology data network. https://toxnet.nlm.nih.gov. Accessed Oct 2017

USEPA (2007) Framework for metals risk assessment. Office of the Science Advisor, Washington. http://www.epa.gov/raf/metalsframework/pdfs/metals-risk-assessment-final.pdf. Accessed 28 October 2013

van Leeuwen CJ, Hermens JLM (1996) Risk assessment of chemicals: an introduction. Kluwer, Dordrecht

Weidenhaupt A, Arnold C, Muller SR, Haderlein SB, Schwarzenbach RP (1997) Sorption of organotin biocides to mineral surfaces. Environ Sci Technol 31:2603–2609

Whalen MM, Hariharan S, Loganathan BG (2000) Phenyltin inhibition of the cytotoxic function of human natural killer cells. Environ Res 84:164–169

WHO (1999) Tributyltin oxide. Concise international chemical assessment 14. World Health Organisation, Geneva

Woolley AA (2008) Guide to practical toxicology: evaluation, prediction, and risk.2nd edn. Informa Healthcare, New York

Wynne KJ (1988) An introduction to inorganic and organometallic polymers. ACS Symp Ser 360:1–4

Zhao CM, Campbell P, Wilkinson K (2016) When are metal complexes bioavailable. Environ Chem 13:425–433. https://doi.org/10.1071/EN15205

Chapter 5
Worldwide Distribution of OMCs

Abstract Due to several studies on the acute ecotoxicity of tributyltin (BT) and phenyltin (PhT), these chemicals are prohibited worldwide. It is shown that butyltin (BT) and phenyltin (PhT) are dangerous anthropogenic substances that are producing new problems in aquatic environments. TBT has long been regarded as a historical contaminant, and it was anticipated that its environmental concentration would progressively decrease over time. TBT inputs are still being recorded globally, despite attempts to reduce TBT-related pollution. This chapter monitors the dispersion of OMCs to determine whether these substances are still present in aquatic systems. BT and PhT levels worldwide are represented in the literature studies. Three case studies concerning levels of BT and PhT around the world are represented. An emphasis on the most recent case study issued in 2024 is reported.

Keywords BT-PhT · Apportionment · Worldwide

5.1 An Overview of Several Environmental Problems

Except for the Baltic Marine Environment Protection Commission, the Convention for the Protection of the Marine Environment of the North-East Atlantic, and the Arctic Monitoring and Assessment Programme regions, most regions lack coastal observations and assessments of trends; as a result, the focus is on the coasts of Europe as well as the North Atlantic and Arctic regions. Global fish catch demonstrates that every location produces at least certain animals at a higher trophic level, surpassing suggested levels, and so, all oceans are impacted regions. In conclusion, cadmium, lead, and mercury are still present at common biota concentrations above baseline levels, with variations in both space and time. There is still strain on predators, with metal levels as a contributing element. The World Mineral Statistics database indicates that over the past 10 years, the annual global output of Cd has been comparatively constant at between 21,000 and 26,000 tons (Brown et al. 2019), while production peaked between 2014 and 2017. Since the peak output of 5.3×10^5 tons of lead annually during 2013–2014, the mine production has dropped by over

T. O. Said, G. M. El Zokm, *Organometallic Compounds in the Marine Environment*, https://doi.org/10.1007/978-3-031-99189-9_5

10%. Over the same period, the production of refined lead has been quite consistent at about 11×10^6 tons. China alone is in charge of producing over half of the world's lead annually.

One of the main factors influencing how these elements are processed in the environment is the biochemical modification of Hg, Sn, As, Sb, Bi, Se, and Te through the formation of volatile metal hydrides and alkylated species (volatile and involatile). Such species can lead to bioaccumulation in lipophilic environments and, in most cases, increase the element's environmental mobility. The more lipid-soluble organometals can be quite poisonous, even though the inorganic versions of the majority of these substances, such as arsenic and mercury, are well characterized, and some of them show modest toxicity, such as tin and bismuth. Methylmercury toxicity and tumor development in rats after exposure to tributyltin oxide or dimethyl arsinic acid are two examples (Dopp et al. 2004). On the other hand, little is known about the cellular mechanisms of action and genotoxicity (and neurotoxicity) of OMC compounds. Numerous studies have shown that the formation of such OMC species can and most likely will occur when anaerobic conditions (at least on a microscale) are coupled with readily available metal(loid)s and methyl donors in the presence of suitable organisms. Both man-made environmental systems (such as sewage treatment facilities and waste disposal sites) and natural ecosystems (such as wetlands and pond sediments) can have such anaerobic conditions. Under aerobic circumstances, some methylation can also occur. An overview of the environmental dispersion of OMC compounds and their potentially harmful effects on human and animal health is provided in this article, specifically genotoxic effects in vitro and in vivo. Lead, cadmium, arsenic, mercury, and other toxic metal (loids) cause several environmental problems and can harm people if they build up in large enough amounts. Because of biological metabolism or physicochemical modification that results in redox shifts, the organo-forms of certain inorganic metal(loid)s are believed to be more dangerous to humans than their elemental equivalents, even though they have not been thoroughly studied. The organic forms of several metal (loids), such as Hg, As, Pb, Sn, Sb, Se, and Cd, have been studied for their ecotoxicity. Additionally treated in detail were the transition, speciation, and transformation mechanisms of these hazardous metal(loid)s in soil-water-plant-microbial systems (Majumdar et al. 2024). The status of harmful OMCs has been highlighted in this review, which is necessary to alert the scientific community to the urgent issue of the dispersion of OMCs in the aquatic environment. Given the increasing dumping and piling of OMCs in the environment, this evaluation highlights the necessity of a thorough review of the past-present status with possible remediation approaches.

To determine the importance of this class of chemicals for human health, the physicochemical and biological generation, distribution, and transformation processes are examined after OMCs in substantial biogenic and anthropogenic deposits have been profiled. By estimating the global levels of biogenic and anthropogenic emissions of OMCs, multidisciplinary articles authored by experts from a variety of disciplines, including biogeochemistry, ecotoxicology, analytical chemistry, microbiology, and genetics, provide insight into processes that affect the genesis, distribution, and stability of OMC species as well as their interactions with other matrix compounds and one another. The authors assess environmentally significant sources from a toxicological perspective to find possible “hot spots” of OMCs that might harm ecosystems.

5.2 Quick Survey on OTCs: Sediment as a Scavenger for Pollutants

Human activity has rapidly increased in densely populated places during the past few decades, increasing the risk of environmental degradation, especially in coastal systems (Clark 2001; Usero et al. 2005; Zoller 2006). Numerous pollutants are released by industry, agriculture, urban wastewater treatment, port operations, and tourism, and these contaminants combine with the environment’s natural components (Kucuksezgin et al. 2006). The coastal zone is ultimately reached by both natural and man-made inputs, primarily through river transport. However, dust and artificial aerosols can diffuse and deposit in the atmosphere on the sea-atmosphere interface. The survey of particulate discharges into sensitive littorals is crucial, as is the fact that sediments serve as a preferential sink for pollutants (Roussiez et al. 2006).

Sediment from El-Mex Bay, Abu-Qir Bay, Eastern Harbour, and Western Harbour—all semi-closed locations of Alexandria, the Egyptian Mediterranean—was used to test OTCs. Due to business activities within the harbor and the influence of garbage discharged from the El Nubaria canal, the Western Harbour has a maximum of 6.34 μg g^{-1} for total tin $\sum$(DBT, TBT, and DPT), compared to 1.63, 0.33, and 1.06 μg g^{-1} in other areas (Shreadah et al. 2006).

They recorded that TBT was dominant and showed the maximum level compared to DBT and DPhT in all sampling areas. They also declared that TBT precipitation into sediments decreases the degradation rate of its derivatives. Abdel Ghani et al. (2010) examined BTCs in the sediment of 39 stations spread out along Egypt’s Mediterranean Sea using GC-ECD. GAAS was used to determine the total tin concentrations. A mean concentration of 2.29 and 3.74 μg g^{-1} was recorded for the eastern and western sectors, respectively. At the majority of stations, DBT was not found. For the eastern and western sectors, the TBT percentages ranged from 0.87 to 55% and from 0.17 to 18%, respectively. This might be related to the western sector’s heavy shipment activities, where TBT paints were used as antifouling agents. They reported that TBT content exceeded the maximum trigger value (70 mg g^{-1}), which could endanger the benthic biota in the area. Also, TBT concentration was used to classify the sediments as extremely severely polluted. OTCs and OLCs were examined in the sediments of the Gulf of Suez (Shreadah et al. 2011). TBT concentrations varied from 0.17 to 1.53 μg g^{-1} in sediment samples, primarily influenced by shipping activity; however, DBT varied from 0.054 to 1.026 μg g^{-1}. TBT and DBT were found to have a significant correlation with $r = 0.64$ at $p = 0.05$, suggesting that the occurrence of DBT is primarily related to TBT degradation.

After carefully examining the DPhT data in this study, it was determined that the maximum concentration of 0.95 μg g^{-1} was recorded for DPhT. This value was attributed to the usage of TPhT as an antifouling agent, sewage discharge, and industrial discharge. OLCs ranged from 8.92 to 303 ng g^{-1}. The results, which demonstrated a notable setting of OLC into sediments, are due to the adsorption of OLC on suspended particles and sediments in the water column. In addition to spelling and direct evaporation, other potential sources of OLC pollution include incomplete combustion of fuels released by automobile engines, which is why the maximum concentrations of OLC were recorded near Suez City's highway roads. Water circulation may also have an impact on the OLC transport mechanism, which results in increased pollution on the western side of the gulf from a persistent anticlockwise circulation in Suez Bay (the northern part of the Suez Gulf). Noubissie et al. (2017) used GC/ICP/MS to analyze organic forms of Hg, S, and Pb in market garden soil of Ngaoundere, Cameroon. After treating samples of both amended and unamended soils, they showed that MeHg was present in the soil at various concentrations, from non-detected in non-amended soil to 172 μg kg^{-1} in amended soil. With concentrations ranging from 18.45 to 44.61 μg kg^{-1}, TEtPb was the most prevalent OLC found in the soil. OTC has also been found to contain higher concentrations of MBT, TMT, and DMT. Soil contamination by OMC, OTC, and OLC is significantly influenced by soil treatments of combustion debris. A review of the entire problem's investigation will aid in future developments to address the issue for the benefit of marine and human life. Some of the pollution issues that Japan has faced in the past and is still facing are listed here. These include radioactive contamination from the nuclear power plant explosions caused by the large tsunami, low-concentration pollution issues like dioxins and endocrine disruptors, and Minamata disease (Tanaka et al. 2023). One of these is the Hg poisoning of waterways due to Indonesia's small-scale gold mining activities, according to the researchers that conducted a collaborative survey of the pollution (Djuwita et al. 2021). In 2001 and 2002, they conducted surveys on gold mining in the Java and Central Kalimantan rivers and measured the levels of Hg in river water and sediments (Sumartias 2020). In 2017 and 2018, research was done on OTC and heavy metal contamination in bottom sediment from three important locations in the Eastern Gulf of Finland, Baltic Sea (Zhakovskaya et al. 2022). ∑OTCs ranging from 5.7 to 123 ng g^{-1} exceed international standards, potentially causing toxic effects and pathological changes in aquatic organisms. In 2017, TBT was the most prevalent component of OTCs in bottom sediments. Its high content average 6.3 and 30 ng g^{-1} for Moschny and Gogland islands, respectively, and low degradation coefficient (BDI: 0.1–0.8) indicate a long-term source of pollution. The 2018 samples show a BDI increase of 1.7–3.3, indicating TBT degradation over the year. Because environmental contamination can result in permanent changes to the environment, it is essential to monitor it in research areas. In September 2014, the distribution of booster biocides and antifouling organotin compounds (OTCs) in the surface sediments of Samsun Port (Black Sea, Turkey) was investigated using gas chromatography-tandem mass spectrometry (GC/MS/MS) (Çetintürk and Ünlü 2022). The concentrations of OTCs varied between <1.0 and 669.6 ng g^{-1}, dw. Diuron with a range from <1.0 to

11.28 ng g^{-1} was found in 70.58% of the sediments under investigation among the studied booster biocides, whereas Irgarol was found in two stations ranged from <1.0–26.53 ng g^{-1}. The port's park and repair areas contained high levels of Irgarol and traces of recently input OTC. TOC and sediment characteristic types were the primary relevant parameters in the antifouling accumulation in the port area. This port was a highly polluted area compared to various environmental sediment quality criteria. The OTC and diuron concentrations could serve as a baseline reference level for upcoming monitoring initiatives in Turkish ports. In 2019, research was conducted on the presence of OTC pollution in benthic ecosystem components in the Barents Sea (South Shore, Kola Peninsula) and the Fram Strait (Icefjord, Svalbard Archipelago). The surface sediments were used to measure six species of OTCs, including tributyltin, dibutyltin, tetrabutyltin, triphenyltin, and tri-cyclo-hexyl-tin (Metelkova et al. 2022). The obtained results indicated a detectable level of OT contamination in the studied samples. The total concentration of six tin compounds ranged from 35.0 to 139 ng g^{-1}. The levels of tributyltin found in bottom sediments did not surpass the internationally set limits. The degradation indices of BT derivatives showed that tetrabutyltin and TBT underwent active transformation in bottom sediments. OTC concentrations in the sediment sampled for that study were comparable to those in other Arctic areas. Bandara et al. (2021) determined TBT contamination in Sri Lanka's sediment. TBT concentrations ranged from 107 ± 4.10 ng kg^{-1} to 17.0 ± 1.40 ng kg^{-1}. The Colombo Port recorded the highest TBT level in sediment. BTC and PhTC residues were measured in marine sediment cores collected from the estuarine waters between Sweden and Denmark (Jacobsen and Strand 2000). TBT ranged from the low detection limit <1.00 pg kg^{-1} to 24.0 pg kg^{-1} dw found in the cores. To make comparing butyltin species easier, all OTC concentrations are normalized to (in atoms (Sn)). The TBT degradation products, DBT and MBT, were only found in trace amounts. According to first-order TBT degradation rates in sediment, estimated half-lives were 8.80 and may be due to low degradation rates in sediments. For the past ten years, there has been no TPhT or degradation products. The low predominance of degradation products indicates an even slower degradation process. Surface sediments show signs of diminishing TBT concentrations, most likely due to legislation and a decrease in the use of TBT by boats beneath 25.0 m.

5.3 Hot Spots in Aquatic Systems

This chapter thoroughly examines numerous case studies with a focus on hot spots in aquatic systems. OTCs were first commercially employed in the plastics industry in the 1940s. To this day, OTCs are used as an addition to stabilize polyvinyl chloride (PVC) and protect it against the effects of light and heat. When over-the-counter (OTC) drugs' biocidal qualities were found in the latter part of the 1950s, they were used as pesticides, fungicides, molluscicides, and wood preservatives in textiles and various water systems. The distribution of organic forms of Hg, As, Sn, Sb, Bi, Cu, nickel, and lead compounds (OMCs) is monitored in this chapter (Dopp et al. 2004; Kagilev et al. 2024). It has been established that the primary source of TBT contamination is the marine sector. TBT was most concentrated in shipyards and nearby ports and marinas (Kucharski et al. 2022). It is the wind that carries ship-cleaning dust into the marine environment. TBT leaching rates can reach up to 6 μg cm^{-2} day^{-1} when freshly painted, according to estimates of the leaching rate. Numerous researchers (Said et al. 2006; Shreadah et al. 2006, 2008; Batista-Andrade et al. 2018; Younis 2020; Lee et al. 2022; Gomez et al. 2024) have confirmed that TBT, one of the most dangerous OMCs, is still present in marine ecosystems. There are two primary causes for this. First, the TBT sedimentation is facilitated by its hydrophobicity and extended half-life periods. The resuspended sediments may then release or diffuse TBT into the water column, where biota would be able to access it. Second, because of a lack of regulation, antifouling paints containing OTCs are nevertheless utilized in some regions of the world even though they are illegal. Furthermore, those paints containing TBT are still sold for use in nations that have not signed the agreement or in places with lax laws (Uc-Peraza et al. 2022). Because of seasonal fluctuations and their distinct characteristics, such as salinity, organic matter content, oxygen content, and the initial concentration of TBT, OTCs decay more slowly in sediments but more quickly in seawater (Abraham et al. 2017). Furthermore, biodegradation is the only process that occurs in sediments (Lee et al. 2022). There are over 5000 species of marine crustaceans, including numerous bioindicator species concerned with the effects of TBT (Verslycke et al. 2005; Parmentier et al. 2019). For example, McClellan-Green et al. (2007) used crustacean bioindicator/bioassay species, including copepods like *Amphiascus tenuiremis*, *Tisbe bulbisetosa*, and *Acartia tonsa*, as well as the grass shrimp *Palaemon pugio*, as particularly relevant bioindicator species in TBT toxicokinetics and TBT effects on toxicological end points such as development, morphology, behavior, and survival. TBT poisoning (Vogt et al. 2018). Kusk and Petersen (1997) discovered that *Acartia tonsa* larval development is highly sensitive to TBT, with inhibition of larval development occurring in a standard 48-hour biotest already at a TBT concentration of 1 ng L^{-1}, demonstrating a sensitivity similar to imposex in *N. lapillus*. In addition to ecdysteroids, juvenile hormone and methyl farnesoate (MF), two sesquiterpenoids, are crucial for crustacean development and transformation. TBT has been demonstrated in crustaceans to reduce juvenile growth rates (Dahllof et al. 2001), inhibit larval developmental ratios (Waldock et al. 1999), impair reproductive performance (IPCS 1990), and alter the local community structure (Aono and Takeuchi 2008). By interfering with methyl farnesoate and ecdysteroid signaling through interactions with other nuclear receptors, TBT can impact sexual maturity and growth. By disrupting the cytochrome P450 system-mediated steroid metabolism, it can also have an impact on reproduction (Vogt et al. 2018). TBT can influence carbohydrate and lipid homeostasis by interacting with RXR and CHH signaling. One species that may be especially pertinent for evaluating the impacts of

5.3.7 *North Pacific Ocean*

The burning of biomass and fossil fuels causes significant seasonal and episodic pulses in the inputs from the Asian continent to the East China Sea and the North Pacific (Qin et al. 2016). Total Hg levels in the deep waters of the North Pacific are higher than those in the surface and intermediate waters, despite comparisons with historical data showing that concentrations have not changed over the past 20 years (Munson et al. 2015).

5.3.8 *South Pacific Ocean*

A thorough analysis of the South Pacific's mercury distribution revealed notable methylmercury concentrations of up to 20% of total mercury and high concentrations in the Peruvian upwelling zone (Bowman et al. 2016). Although the region's data is insufficient to determine trends since the first assessment, the numbers seem constant. Although there is a net source of mercury entering the atmosphere from the tropical South Pacific, the exchange flux is less than that of the North Atlantic (Mason et al. 2017).

5.3.9 *Southern Ocean*

The Southern Ocean's total mercury concentrations are similar to those of the Atlantic and South Pacific Oceans. Nonetheless, certain unique regional characteristics exist, such as methylmercury formation south of the southern polar front, mercury enrichment in brine during sea ice development, and net mercury deposition along the Antarctic sea ice's ice edge (Cossa et al. 2011). Despite its distant location, the lead contents in the water (6.2 μg/l) are equivalent to those seen in more industrialized areas, such the Baltic Sea (Schlosser et al. 2016).There aren't enough metal data points in the area to identify any patterns since the initial assessment.

5.4 Micropollutants' Global Spread, both Qualitatively and Quantitatively

Micropollutants (MPs) are a broad category of biological disruptors that, at extremely low doses (<1 μg L^{-1}), can be hazardous to both people and wildlife. Marine life is impacted by these mostly man-made contaminants, which have been found in many parts of the world, including the deep sea. Because deep-sea MPs

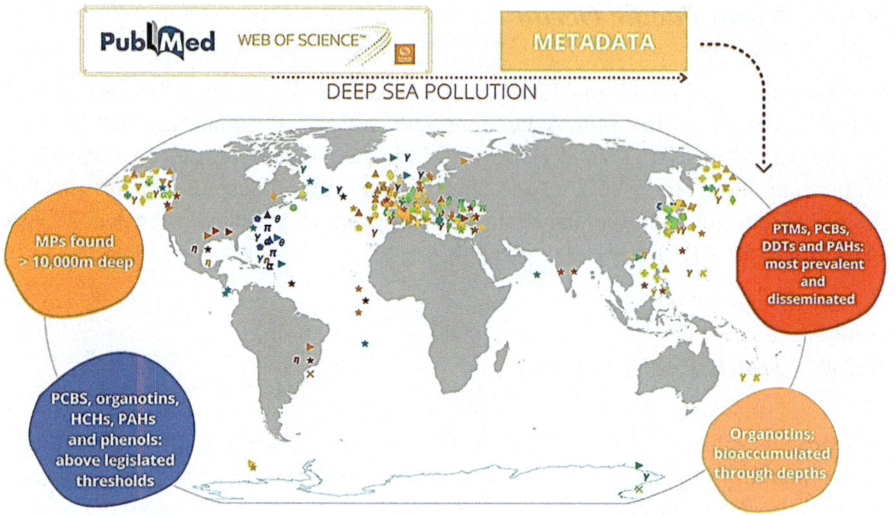

Fig. 5.2 Global distribution of MPs in the deep sea (Spindola Vilela et al. 2022)

may be harmful, Fig. 5.2 shows their global distribution, quantity, occurrence, and possible effects (Spindola Vilela et al. 2022).

The findings demonstrated that MPs are more commonly recorded in the Northern Hemisphere, where they are found in larger numbers and have reached several oceanic zones. In addition to being more commonly researched at depths up to 3000 m, MPs may also be found at depths deeper than 10,000 m. At a depth of ≥200 m, the most common and extensively dispersed MPs were found to include potentially toxic metals (PTMs), polychlorinated biphenyls (PCBs), dichlorodiphenyltrichloroethane (DDTs), OTCs, and polycyclic aromatic hydrocarbons (PAHs). PTMs are found in large quantities in the deep sea, up to 3000 m below the surface. Aluminum is the most common PTM, followed by copper and zinc. In both species and environmental samples, PCBs, OTCs, hexachlorocyclohexanes (HCHs), PAHs, and phenols were found to have accumulated over legally mandated limits or established toxicity levels. According to our analysis, MPs can be seen as sinking in the deep sea (Spindola Vilela et al. 2022).

Human security and health depend on the condition of the environment (WHO 2010). However, the world's growing population, fast industrialization, and urbanization have led to several anthropogenic pollutants becoming more common in the biosphere (Welsch 2006; Khan and Ghouri 2011; Villarrubia-Gómez et al. 2018; Yadav et al. 2021; Zandalinas et al. 2021), putting human health and the environment at risk. Activities that produce and/or dispose of pollutants of both natural and man-made (xenobiotic) origin include deforestation, burning, industrial emissions, agriculture, pharmaceutical and chemical synthesis, and home and industrial wastewater (Hill 2010). The removal of contaminants and organic matter is mostly dependent on wastewater treatment facilities (Olyaei et al. 2018; Panico et al. 2013). The environment, including deep-sea regions, is contaminated by many pollutants and toxic compounds that cannot be fully eliminated by traditional effluent treatments, despite their effectiveness for many substances (Gothwal and Shashidhar 2015; Lammel et al. 2015; Combi et al. 2016; Stamm et al. 2016; Lambert 2018; Pereda et al. 2020). Other variables influencing aquatic pollution include storms and land runoff (Holt 2000; Rippy et al. 2017; Phillips et al. 2018). The most commonly detected groups of pollutants in environmental samples are several MPs, such as PCBs, aliphatic and aromatic hydrocarbons, OTCs, surfactants, PTMs, and microplastics (Ivar do Sul and Costa 2014; Singh et al. 2019). According to Stamm et al. (2016) and Emadian et al. (2021), MPs are mostly man-made pollutants that are present in the environment at comparatively low quantities (ng L^{-1} to μg L^{-1}). MPs may have detectable impacts on biological targets even at such low concentrations (Vilela et al. 2018). According to Stamm et al. (2016), the threshold concentration that separates macro- from micro-pollutants is typically about 1 μg L^{-1}. Phosphorus, dissolved organic carbon, and nitrate are macro-pollutants (i.e., typically found in quantities over 1 μg L^{-1}), whereas substances like estrogens, pesticides, and medications that are frequently found at concentrations <1 μg L^{-1} are consequently regarded as MPs (Stamm et al. 2016). PTMs can be classified into either group since they can exist in a broad range of quantities, from ng L^{-1} to μg L^{-1} (Stamm et al. 2016; Pourret and Hursthouse 2019). Various aquatic systems include various MP classes, which bioaccumulate in various animals (Lambert 2018). The kinetics of MP transformation, which dictates the level of water pollution and MP consequences, may be influenced by a variety of parameters, including biodegradability, sorption, photodegradation, volatilization, and environmental factors (e.g., temperature, solar radiation) (Schwientek et al. 2016). The scientific community is very concerned about synthetic and xenobiotic compounds because of their potential for toxicity and bioaccumulation, as well as their negative effects on humans and animals (Schwarzenbach et al. 2006; Sarkar et al. 2016; Lambert 2018).

Water may carry MPs in the aquatic environment to far-flung locations like the deep sea (Borghi and Porte 2002; Van Cauwenberghe et al. 2013; Pulster et al. 2020). The exploitation of natural resources (oil, gas, fish), climate change (acidification, global warming), and waste pollution are the main threats to the deeper ocean regions, which make up roughly 75% of the global biosphere (Ahnert and Borowski 2000; Ramu et al. 2006; Ramirez-Llodra et al. 2011; Gleckler et al. 2016; Danovaro et al. 2017). Deep-sea regions may become sinks and reservoirs of persistent pollutants due to abiotic factors like lack of light, low temperatures, and high pressures that hinder the breakdown and removal of contaminants (Woodwell et al. 1971; Oshihoi et al. 2009; Takahashi et al. 2014; Woodall et al. 2014; Courtene-Jones et al. 2017; Dasgupta et al. 2021). Whether abiotic factors in the deep sea can make pollutants more or less harmful is not clear (Mestre et al. 2014). Furthermore, there is no qualitative and quantitative assessment and summary of the extent and distribution of MPs in this remote biome, despite the numerous reports indicating the presence of a variety of MPs in the deep sea (Borghi and Porte 2002; Kádár et al. 2007; White et al. 2012; Courtene-Jones et al. 2017; Romero-Romero et al. 2017;

Munschy et al. 2019). These pollutants are common in marine ecosystems at depths of 200–10,000 m, according to the qualitative and quantitative mapping of MPs in the deep sea. The most common and extensively spread MPs are PTMs, particularly in the Northern Hemisphere and the oceans of Europe. PCBs, DDT, OTCs, and PAHs are next in line. PAHs have mostly been found in the Gulf of Mexico, perhaps the Deepwater Horizon disaster that struck this area in 2010 and the ensuing considerable local efforts to look into the accident's effects. MPs were discovered to be deposited in sediments, saltwater, and living things. HCHs, PCBs, OTCs, PAHs, and phenols were found in concentrations. Over their MAC-EQS (maximum acceptable concentrations). Also highlighting the concerning finding that MPs (PCBs and PBDEs) are nearly universal in our oceans following their discovery at depths above 10,000 meters in the Mariana and Kermadec trenches in the Pacific Ocean. Since microplastics are widely distributed and have been found in the biota and environment at depths of up to 6000 m, deep-sea ecosystems also appear to be a hot spot for microplastic pollution. As a result, they may qualify for inclusion on the World Health Organization's priority list of environmental and health hazards. These results suggest that in addition to using and enhancing MP mitigation strategies, monitoring the presence and toxicological effects of MPs in deep-sea locations is necessary.

Another class of EDCs that can bioaccumulate in the food chain are organotins, which are often found in the samples. These can cause imposex (i.e., females acquiring male structures) or affect penis size, oocyte development, and consequently the reproduction of marine organisms (Casals-Casas and Desvergne 2011; Rumampuk et al. 2018). Numerous studies have documented the detrimental effects of tributyltin (TBT), a general term for a class of organotins commonly found due to their use in antifouling paint (Gibbs et al. 1991; Bettin et al. 1996; Wirzinger et al. 2007; Van den Broeck et al. 2009; Harino et al. 2010; Peranandam et al. 2014; Harino 2016; Matthiessen 2019; Zhang et al. 2021).

5.5 Case Studies

Alkyl tin compounds (ATC) and alkyl lead compounds (APC) were determined in the Gulf of Suez, Egypt, using GC/FID and GAAS (Said et al. 2010). Furthermore, ICP/OES was used to determine the total Sn and Pb concentrations relative to their respective percentages in the total metal content. OTCs were reported to be high, ranged from 44.2 to 82.4 ng l^{-1}. This is because of major input sources and shipping activity in the gulf region. OTCs were higher than the lowest toxicity thresholds (2.00 ng l^{-1}) for more delicate organisms. OLCs reached their peak values in the middle of the gulf, which is regarded as a major hub for the extraction and transportation of oil from numerous offshore wells. In addition, APC concentrations were far below the danger threshold (100 μg l^{-1}).

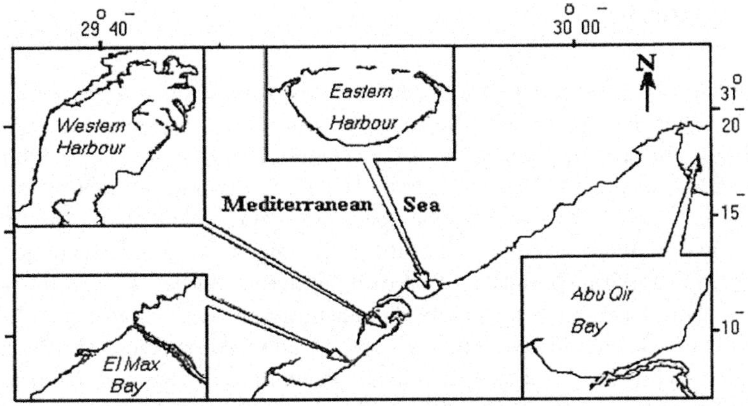

Fig. 5.3 Sampling locations, Alexandria, Egypt (Said et al. 2006)

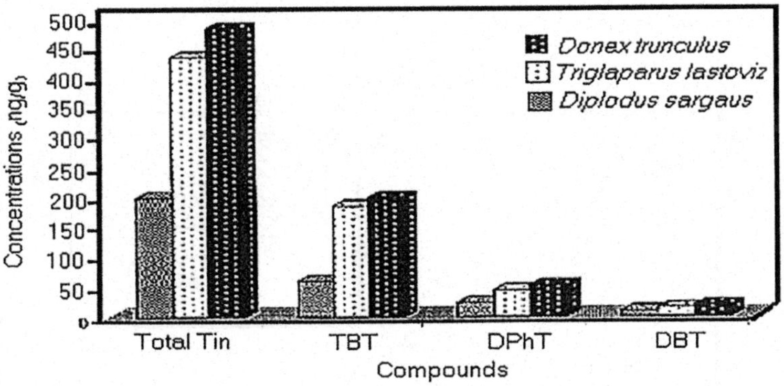

Fig. 5.4 The average concentrations of different forms of tin in three different biota samples (Said et al. 2006)

5.5.1 Case Study 1

Said et al. (2006) recorded the residual concentration of OTCs in different demersal and pelagic fish and bivalves from Alexandria, Egypt (Fig. 5.3). In general, the concentrations of TBT were the highest among all biota samples compared to DBT and DPhT (Fig. 5.4). This trend might be due to the worldwide use of TBT as biocides in antifouling paints and as a preserving agent for wood, fungicides in agricultural activities, and heat and UV stabilizer for PVC. When TBT is applied, it is directly released into the water body, where it accumulates in aquatic life and slows the pace at which it breaks down into derivatives.

5.5.2 *Case Study 2*

Said et al. (2010) investigated alkyl tin and alkyl lead compounds in the coastal water over the Suez Gulf. Significant input sources and shipping activities in the gulf region are primarily to blame for the high quantities of OTC, which ranged from 44.2 to 82.4 ng L^{-1}. The lowest toxicity criteria for more susceptible organisms (2.00 ng L^{-1}) were surpassed by the spectrum of OTC concentrations in surface water. However, the central portion of the gulf, which is regarded as a major hub for the extraction and transportation of oil from numerous oil fields, including offshore wells, is where OLC reaches its highest concentrations. In the majority of locations, Et_4Pb was not detectable. Nonetheless, among alkyllead species, Me_4Pb predominated. The findings showed that alkyl lead concentrations were still much below the 100 µg/l danger level (Fig. 5.5). Generally, the highest OLC values found in the Suez Gulf's water were 198.37, 728.09, 1751.59, and 53.24 ng/l for Et_4Pb, Me_4Pb, Et_3Pb^+, and Me_3Pb^+, respectively. Stations VI, VII, and VIII (the middle portion of

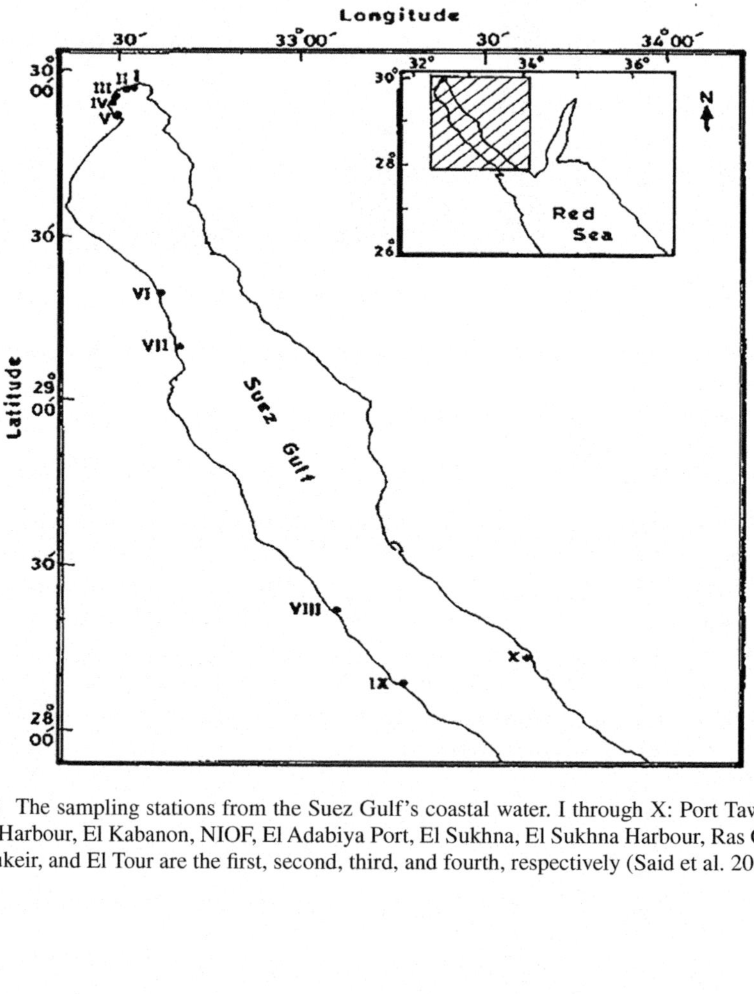

Fig. 5.5 The sampling stations from the Suez Gulf's coastal water. I through X: Port Tawfiq, El Zaitiya Harbour, El Kabanon, NIOF, El Adabiya Port, El Sukhna, El Sukhna Harbour, Ras Gharib, Ras Shukeir, and El Tour are the first, second, third, and fourth, respectively (Said et al. 2010)

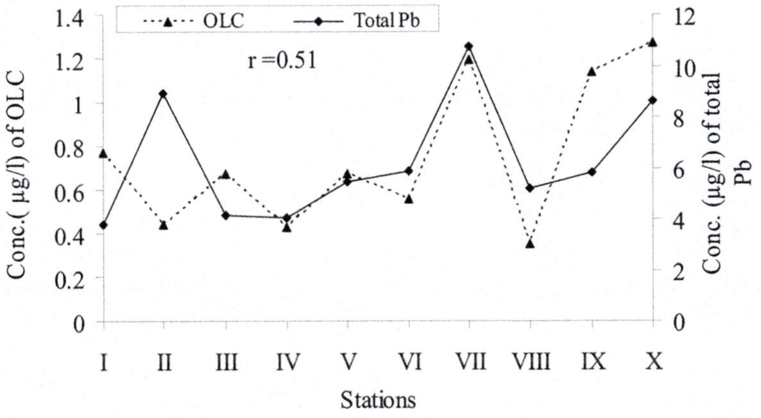

Fig. 5.6 The correlation between OLC and total lead concentrations at various Suez Gulf stations ($r = 0.51, \rho = 0.05$) (Said et al. 2010)

the gulf), which are regarded as major hubs for the extraction and transportation of oil from several oil fields, including offshore wells, were where these figures were recorded. Because all of the Pb is in the form of Et_4Pb, a sizable portion of OLC originates from non-combustion gasoline that is discharged into the gulf from coastal gas stations and during maritime activities. A negligible correlation of $r = 0.51$ ($\rho = 0.05$, $n = 8$) was found between the average concentrations of OLC and inorganic Pb (Fig. 5.6).

5.5.3 *Case Study 3: Lower Yangtze River in China*

Sun et al. (2024) determined butyltin and phenyltin concentrations in surface water and sediments of the Yangtze River in 2022–2023. They analyzed their occurrence and sources and assessed their ecological risks. The findings indicated that butyltin and phenyltin compounds are widespread. The total concentration of OTCs in surface water was as high as 42.41 ng Sn l^{-1}, with an average of 25.71 ng Sn l^{-1}, and the total concentration of OTCs in sediments reached up to 111.5 ng Sn g^{-1} dw, averaging 61.1 ng Sn g^{-1} dw (Fig. 5.7).

The ecological risk of OTCs in the surface water and sediments of the Yangtze River was assessed (Fig. 5.8). Sun et al. (2024) indicated that the ecological risks associated with MPhT, MBT, DBT, and DPhT were relatively low. However, TBT and TPhT pose high ecological risks. TBT demonstrated RQ > 1, at several monitoring sites, signifying concentrations that meet or surpass the ecological risk thresholds, with significant biotoxic effects that could impact aquatic life even at minimal concentrations. TPhT similarly exhibited high risks at certain locations, particularly near industrial parks, highlighting its potential as a pollutant and underscoring the importance of mitigating external inputs. The pollution problem of TBT in Yangtze

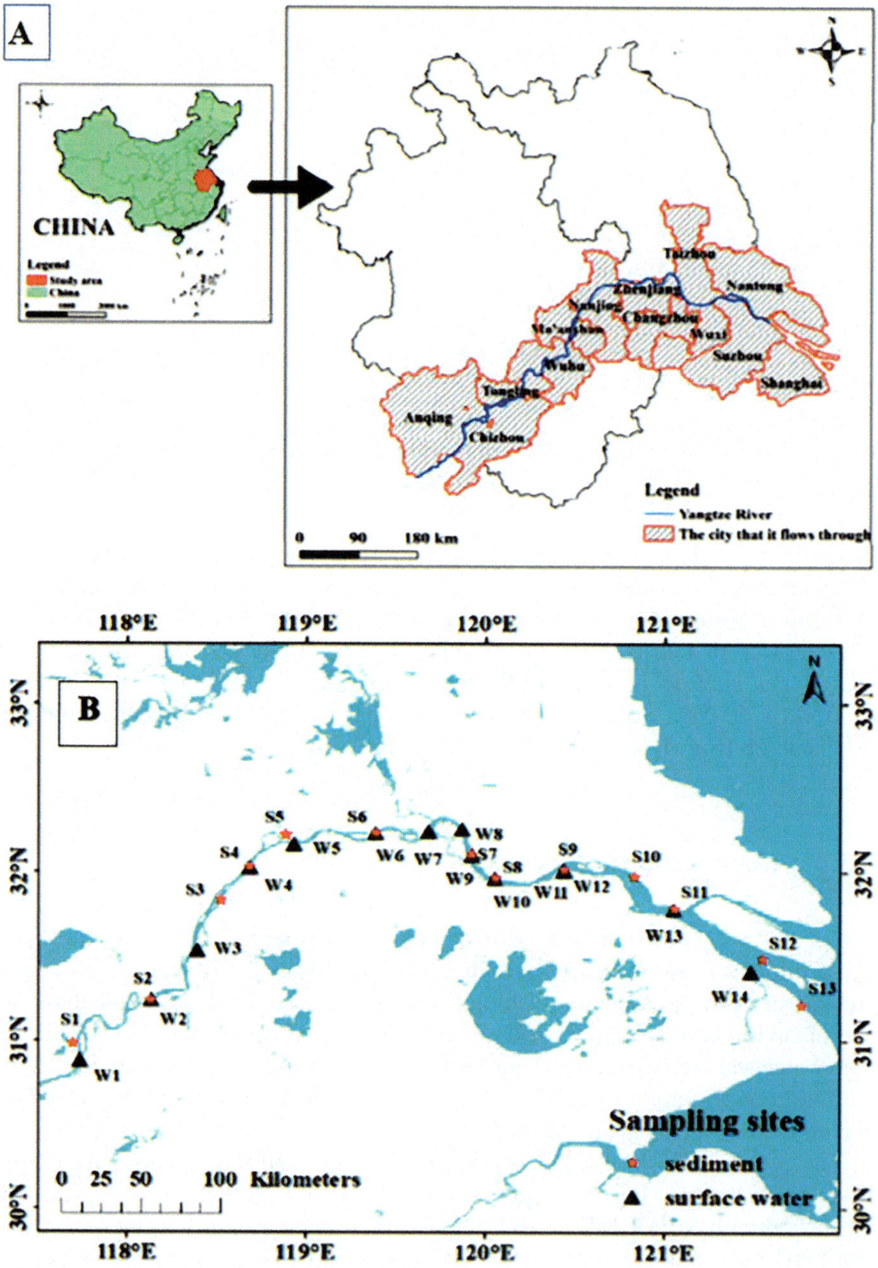

Fig. 5.7 (**a**) The study area, (**b**) The sampling sites (triangles denote the location of surface water sampling, and stars indicate the site for sediment sampling) along the lower Yangtze River (Sun et al. 2024)

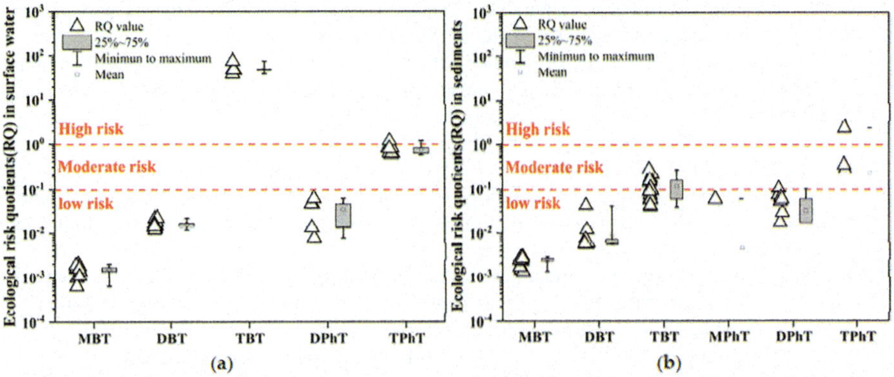

Fig. 5.8 The ecological risk of organotin compounds in the surface water and sediments of the Yangtze River (Sun et al. 2024). Two-panel figure showing bar charts of ecological risk quotients (RQ) for various compounds in surface water and sediments. Panel (**a**) displays RQ values for MBT, DBT, TBT, DPhT, and TPhT in surface water, while panel (**b**) shows the same for sediments. Risk levels are indicated with dashed lines: low risk, moderate risk, and high risk. Symbols represent RQ value, 25%-75% range, minimum to maximum, and mean. The y-axis is logarithmic, ranging from 10^{-4} to 10^{2}.

River should not be ignored. TBT stems from human introduction and is also one of the most toxic materials in an aquatic environment. Even at a lower level of concentration, it can trigger a multitude of adverse effects, especially in sediments that can also persist for a long time. Therefore, OTCs in Yangtze River still present challenges in its governance.

References

Abdel Ghani SA, Shobier AH, Said TO, Shreadah MA (2010) Organotin compounds in Egyptian Mediterranean sediments. Egypt J Aquat Res 36(3):357–367

Abraham M, Westphal L, Hand I, Lerz A, Jeschek J, Bunke D, Leipe T, Schulz-Bull D (2017) TBT and its metabolites in sediments: survey at a German coastal site and the Central Baltic Sea. Mar Pollut Bull. https://doi.org/10.1016/j.marpolbul.2017.06.020

Ahnert A, Borowski C (2000) Environmental risk assessment of anthropogenic activity in the deep-sea. J Aquat Ecosyst Stress Recov 70:299–315

Al-Ansari EMAS, Abdel-Moati MAR, Yigiterhan O, Al-Maslamani I, Soliman Y, Rowe GT, Wade TL, Al-Shaikh IM, Helmi A, Kuklyte L, Chatting M, Al-Yafei MA (2017) Mercury accumulation in Lethrinus nebulosus from the marine waters of the Qatar EEZ. Mar Pollut Bull 121(1-2):143–153. https://doi.org/10.1016/j.marpolbul.2017.04.024

Aono A, Takeuchi I (2008) Effects of tributyltin at concentrations below ambient levels in seawater on Caprella danilevskii (Crustacea: Amphipoda: caprellidae). Mar Pollut Bull 57:515–523

Bandara KRV, Chinthakac SDM, Yasawardened SG, Managea PM (2021) Modified, optimized method of determination of tributyltin (TBT) contamination in coastal water, sediment and biota in Sri Lanka. Mar Pollut Bull 166:112202

Batista-Andrade JA, Caldas SS, Batista RM (2018) From TBT to booster biocides: levels and impacts of antifouling along coastal areas of Panama. Environ Pollut 234:243–252

Bettin C, Oehlmann J, Stroben E (1996) TBT-induced imposex in marine neogastropods is mediated by an increasing androgen level. Helgol Meeresunters 50:299–317

Bodin N, Lesperance D, Albert R, Hollanda S, Michaud P, Degroote M, Churlaud C, Bustamante P (2017) Trace elements in oceanic pelagic communities in the western Indian Ocean. Chemosphere 174:354–362. https://doi.org/10.1016/j.chemosphere.2017.01.099

Borghi V, Porte C (2002) Organotin pollution in deep-sea fish from the northwestern Mediterranean. Environ Sci Technol 36:4224–4228

Bowman KL, Hammerschmidt CR, Lamborg CH, Swarr GJ, Agather AM (2016) Distribution of mercury species across a zonal section of the eastern tropical South Pacific. Mar Chem 186:156–166

Bridgestock L, van de Flierdt T, Rehkämper M, Paul M, Middag R, Milne A, Lohan MC, Baker AR, Chance R, Khondoker R, Strekopytov S, Humphreys-Williams E, Achterberg EP, Rijkenberg MJA, Gerringa LJA, de Baar HJW (2016) Return of naturally sourced Pb to Atlantic surface waters. Nat Commun 7:12921. (2016). https://doi.org/10.1038/ncomms12921

Brown TJ, Iodine NE, Raycraft ER, Hobbes SF, Shaw RA, Everett P, Kresse C, Deady EA, Bide T (2019) World mineral production 2013–17. British Geological Survey, Nottingham

Casals-Casas C, Desvergne B (2011) Endocrine disruptors: from endocrine to metabolic disruption. Annu Rev Physiol 73:135–162

Çetintürk K, Ünlü S (2022) The first observation of antifouling organotin compounds and booster biocides in sediments from Samsun port area, Black Sea, Turkey. Mar Pollut Bull 176:113408. https://doi.org/10.1016/j.marpolbul.2022.113408

Clark RB (2001) Marine pollution. Oxford University Press, Oxford, p 248

Combi T, Pintado-Herrera MG, Lara-Martin PA, Miserocchi S, Langone L, Guerra R (2016) Distribution and fate of legacy and emerging contaminants along the Adriatic Sea: a comparative study. Environ Pollut 218:1055–1064

Cossa D, Heimbürgera LE, Lannuzel D, Rintoul SR, Butler ECV, Bowieb AR, Averty B, Watsond RJ, Remenyi T (2011) Mercury in the Southern Ocean. Geochim Cosmochim Acta 75(14):4037–4052

Courtene-Jones W, Quinn B, Gary SF, Mogg AO, Narayanaswamy BE (2017) Microplastic pollution identified in deep-sea water and ingested by benthic invertebrates in the Rockall trough, North Atlantic Ocean. Environ Pollut 231:271–280

Cunningham PA, Sullivan EE, Everett KH, Kovach SS, Rajan A, Barber MC (2019) Assessment of metal contamination in Arabian/Persian gulf f ish: a review. Mar Pollut Bull 143:264–283

Dahllof I, Agrenius S, Blanck H, Hall P, Magnusson K, Molander S (2001) The effect of TBT on the structure of a marine sediment community - a boxcosm study. Mar Pollut Bull 42:689–695

Danovaro R, Corinaldesi C, Dell'Anno A, Snelgrove PV (2017) The deep-sea under global change. Curr Biol 27:R461–R465

Dasgupta S, Peng X, Xu H, Ta K, Chen S, Li J, Du M (2021) Deep seafloor plastics as the source and sink of organic pollutants in the northern South China Sea. Sci Total Environ 765:144228

Dastoor AP, Durnford DA (2013) Arctic Ocean: is it a sink or a source of atmospheric mercury? Environ Sci Technol 48(3):1707–1717

Djuwita MR, Hartono DM, Mursidik SS, Budhi TE (2021) Pollution load allocation on water pollution control in the Citarum River. J Eng Technol Sci 53(1):210112. https://doi.org/10.5614/j.eng.technol.sci.2021.53.1.12

Dopp E, Hartmann LM, Florea AM, Rettenmeier AW, Hirner AV (2004) Environmental distribution, analysis, and toxicity of organometal(loid) compounds. Crit Rev Toxicol 34(3):301–333. https://doi.org/10.1080/10408440490270160

Emadian SM, Sefiloglu FO, Balcioglu IA, Tezel U (2021) Identification of core micropollutants of Ergene River and their categorization based on spatiotemporal distribution. Sci Total Environ 758:143656

Esposito M, De Roma A, La Nucara R, Picazio G, Gallo P (2018) Total mercury content in commercial swordfish (Xiphias gladius) from different FAO fishing areas. Chemosphere 197:14–19. https://doi.org/10.1016/j.chemosphere.2018.01.015

Fisher D, Zheng J, Burgess D, Zdanowicz C, Kinnard C, Sharp M, Bourgeois J (2012) Recent melt rates of Canadian Arctic ice caps are the highest in four millennia. Glob Planet Chang 84:3–7

Gibbs PE, Spencer BE, Pascoe PL (1991) The American oyster drill, Urosalpinx cinerea (Gastropoda): evidence of decline in an imposex-affected population (R. Blackwater, Essex). J Mar Biol 71:827–838

Gleckler PJ, Durack PJ, Stouffer RJ, Johnson GC, Forest CE (2016) Industrial-era global ocean heat uptake doubles in recent decades. Nat Clim 6:394–398

Gomez NA, Sturla LJ, Ferrando A, Garrido M, Domini CE (2024) Update on the status of the contamination by organotin compounds in sediment of Nuevo gulf, Argentina. Insights from field and experimental studies. Mar Pollut Bull 200:116087. https://doi.org/10.1016/j.marpolbul

Gothwal R, Shashidhar T (2015) Antibiotic pollution in the environment: a review. Clean 43:479–489

Harino H (2016) Emerging issues on contamination and adverse effects by alternative antifouling paints in the marine environments. In: Biological effects by organotins. Springer, Tokyo, pp 43–70. https://doi.org/10.1007/978-4-431-56451-5_3

Harino H, Iwasaki N, Arai T, Ohji M, Miyazaki N (2010) Accumulation of organotin compounds in the deep-sea environment of Nankai trough, Japan. Arch Environ Contam Toxicol 49:497–503

Hassan H, Elezz AA, Abuasali M, AlSaadi H (2019) Baseline concentrations of mercury species within sediments from Qatar's coastal marine zone. Mar Pollut Bull 142:595–602. https://doi.org/10.1016/j.marpolbul.2019.04.022

Heimbürger LE, Sonke JE, Cossa D, Point D, Lagane C, Laffont L, Galfond BT, Nicolaus M, Rabe B, van der Loeff MR (2015) Shallow methylmercury production in the marginal sea ice zone of the Central Arctic Ocean. Sci Rep 5:10318. https://doi.org/10.1038/srep10318

HELCOM (2018a) Thematic assessment of hazardous substances 2011-2016: supplementary report to the 'state of the Baltic Sea' report. http://stateofthebalticsea.helcom.fi/wp-content/uploads/2018/07/HELCOM_thematic-assessment-of-hazardous-substances-2011-2016_pre-publication.Pdf

HELCOM (2018b) Inputs of hazardous substances to the Baltic Sea. Balt Sea Environ Proc 161:162. www.helcom.fi/Lists/Publications/BSEP162.pdf

Hill MK (2010) Understanding environmental pollution.3rd edn. Cambridge University Press, Cambridge

Holt MS (2000) Sources of chemical contaminants and routes into the freshwater environment. Food Chem Toxicol 38(S21–S27):1683–1692

IPCS (1990) Environmental health criteria for tributyltin compounds, environmental health criteria No. 116. World Health Organization, Geneva, p 277

Ivar do Sul JA, Costa MF (2014) The present and future of microplastic pollution in the marine environment. Environ Pollut 185:352–364

Jacobsen JA, Strand J (2000) Depositional profiles of Butyltin residues in marine sediments from the sound between Sweden and Denmark. Poster session presented at SETAC 21st annual meeting, Nashville, United States

Jonsson S, Andersson A, Nilsson MB, Skyllberg U, Lundberg E, Schaefer JK, Åkerblom S, Björn E (2017) Terrestrial discharges mediate trophic shifts and enhance methylmercury accumulation in estuarine biota. Sci Adv 3(1):e1601239. https://doi.org/10.1126/sciadv.1601239

Kádár E, Costa V, Segonzac M (2007) Trophic influences of metal accumulation in natural pollution laboratories at deep-sea hydrothermal vents of the mid-Atlantic ridge. Sci Total Environ 373:464–472

Kagilev AA, Gafurov ZN, Kantyukov AO et al (2024) The power of in situ spectroelectro chemistry for redox study of organometallic and coordination compounds. J Solid State Electrochem 28:897–910. https://doi.org/10.1007/s10008-023-05765-7

Khan MA, Ghouri AM (2011) Environmental pollution: its effects on life and its remedies. Res World J Arts Sci 2:276–285

Kucharski D, Stasiuk R, Drzewicz P, Skowronek A, Strzelecka A, Mianowicz K, Giebułtowicz J (2022) Fit-for-purpose assessment of QuEChERS LC-MS/MS methods for environmental monitoring of organotin compounds in bottom sediments of the Odra River estuary. Molecules 27:4847. https://doi.org/10.3390/molecules27154847

Kucuksezgin F, Kontas A, Altay O, Uluturhan E, Darılmaz E (2006) Assessment of marine pollution in Izmir Bay: Nutrient, heavy metal and total hydrocarbon concentrations, Environment International 32, Issue 1, Pages 41–51, ISSN 0160-4120, https://doi.org/10.1016/j.envint.2005.04.007.

Kusk KO, Petersen S (1997) Acute and chronic toxicity of tributyltin and linear alkylbenzene sulfonate to the marine copepod Acartia tonsa. Environ Toxicol Chem 16:1629–1633

Lambert JF (2018) Organic pollutant adsorption on clay minerals. In: Schoonheydt R, Johnston CT, Bergaya F (eds) Surface and interface chemistry of clay minerals. Developments in clay science, vol 9. Elsevier, Amsterdam, pp 195–253

Lammel G, Audy O, Besis A, Efstathiou C, Eleftheriadis K, Kohoutek J, Kukučka P, Mulder MD, Přibylová P, Prokeš R, Rusina TP, Samara C, Sofuoglu A, Sofuoglu SC, Taşdemir Y, Vassilatou V, Voutsa D, Vrana B (2015) Air and seawater pollution and air–sea gas exchange of persistent toxic substances in the Aegean Sea: spatial trends of PAHs, PCBs, OCPs and PBDEs. Environ Sci Pollut Res 22:11301–11313

Lee JM, Boyle EA, Toshitaka G, Hajime O, Kazuhiro N, Yolanda E (2015) Impact of anthropogenic Pb and ocean circulation on the recent distribu tion of Pb isotopes in the Indian ocean. Geochim Cosmochim Acta 170:126–144

Lee SH, Chen YS, Chen CF, Albarico FPJB, Lim YC, Wang MH, Chen CW, Dong CD (2022) Butyltin contamination in fishing port sediments after the ban of tributyltin antifouling paint: a case of qianzhen fishing port in Taiwan. Water 14:813. https://doi.org/10.3390/w14050813

Majumdar A, Upadhyay MK, Ojha M, Biswas R, Dey S, Sarkar S, Moulick D, Niazi NK, Rinklebe J, Huang JH, Roychowdhury T (2024) A critical review on the organo-metal(loid)s pollution in the environment: distribution, remediation and risk assessment. Sci Total Environ 951:175531. https://doi.org/10.1016/j.scitotenv.2024.175531

Mason RP, Hammerschmidt CH, Lamborg CH, Bowman KL, Swarr GJ, Shelley RU (2017) The air-sea exchange of mercury in the low latitude Pacific and Atlantic oceans. Deep Sea Res Part I Oceanogr Res Pap 122:17–28

Matthiessen P (2019) The impact of organotin pollution on aquatic invertebrate communities—are molluscs the only group whose populations have been affected? Curr Opin Environ Sci 11:13–20

McClellan-Green P, Romano J, Oberdorster E (2007) Does gender really matter in contaminant exposure? A case study using invertebrate models. Environ Res 104:183–191

Mestre NC, Calado R, Soares AM (2014) Exploitation of deep-sea resources: the urgent need to understand the role of high pressure in the toxicity of chemical pollutants to deep-sea organisms. Environ Pollut 185:369–371

Metelkova L, Zhakovskaya Z, Kukhareva G et al (2022) Organotin compounds [OTs] in surface sediments, bivalves and algae from the Russian coast of the Barents Sea [Kola peninsula] and the Fram Strait [Svalbard archipelago]. Environ Sci Pollut Res 29:34659–34669. https://doi.org/10.1007/s11356-021-18091-0

Munschy C, Chouvelon T, Bely N, Moisan K, Pollono C, Mauffret A, Spitz J (2019) Legacy and emerging organohalogen compounds in deep-sea pelagic organisms from the Bay of Biscay (Northeast Atlantic). Organohalogen Compd 81:108–111

Munson KM, Lamborg CH, Swarr GJ, Saito MA (2015) Mercury species concentrations and fluxes in the central tropical Pacific Ocean. Glob Biogeochem Cycles 29(5):656–676

Noubissie E, Ngassoum MB, Ali A, Castro-Georgi J, Donard OFX (2017) Speciation of organometallic compounds of hg, Sn, and Pb in the market garden soil in Ngaoundere (Cameroon). Euro-Mediterr J Environ Integr 2:18. https://doi.org/10.1007/s41207-017-0028-7

Olyaei MA, Karamouz M, Farmani R (2018) Framework for assessing flood reliability and resilience of wastewater treatment plants. J Environ Eng 144:04018081

Oshihoi T, Isobe T, Takahashi S, Kubodera T, Tanabe S (2009) Contamination status of organohalogen compounds in deep-sea fishes in Northwest Pacific Ocean off-Tohoku, Japan. In: Obayashi Y, Isobe T, Subramanian A, Suzuki S, Tanabe S (eds) Interdisciplinary studies on environmental chemistry. Environmental Research in Asia, TERRAPUB, Tokyo, pp 67–72

OSPAR (2017a) Status and trend for heavy metals (mercury, cadmium, and lead) in fish and shell fish. OSPAR Intermediate Assessment 2017. https://oap.ospar.org/en/ospar-assessments/intermediate-assessment-2017/pressures-human-activities/contaminants/metals-fish-shellfish

OSPAR (2017b) Inputs of mercury, cadmium and lead via water and air to the greater North Sea. OSPAR Intermediate assessment 2017. https://oap.ospar.org/en/ospar-assessments/intermediate-assessment-2017/pressures-human-activities/contaminants/heavy-metal-inputs

OSPAR (2017c) Intermediate assessment 2017. https://oap.ospar.org/en/ospar-assessments/intermediate-assessment-2017

OSPAR (2017d) Status and trend for heavy metals (cadmium, mercury and lead) in sediment. OSPAR intermediate assessment 2017. https://oap.ospar.org/en/ospar-assessments/intermediate-assessment-2017/pressures-human-activities/contaminants/metals-sediment

OSPAR (2017e) Status and trend for heavy metals (mercury, cadmium, and lead) in fish and shell fish. OSPAR intermediate assessment 2017. https://oap.ospar.org/en/ospar-assessments/intermediate-assessment-2017/pressures-human-activities/contaminants/metals-fish-shellfish

Panico A, Lanzano G, Salzano E, Fabbrocino G (2013) Seismic vulnerability of wastewater treatment plants. Chem Eng Trans 32:13–18

Parmentier KFV, Verhaegen Y, De Witt BP, Hoffman S, Delbare DHR, Roose PM, Hylland KDE, Burgeot T, Smagghe GJ, Cooreman K (2019) Tributyltin: a bottom-up regulator of the Crangon crangon population? Front Mar Sci 6:14

Paul M, van de Flierdt T, Rehkämper M, Khondoker R, Weiss D, Lohan MC, Homoky WB (2015) Tracing the Agulhas leakage with lead isotopes. Geophys Res Lett 42(20):8515–8521

Peranandam R, Palanisamy I, Lourdaraj AV, Natesan M, Vimalananthan AP, Thangaiyan S, Perumal A, Muthukalingan K (2014) TBT effects on the development of intersex (ovotestis) in female fresh water prawn Macrobrachium rosenbergii. Biomed Res Int 2014:412619

Pereda O, Solagaistua L, Atristain M, de Gúzman I, Larrañaga A, von Schiller D, Elosegi A (2020) Impact of wastewater effluent pollution on stream functioning: a whole-ecosystem manipulation experiment. Environ Pollut 258:113719

Phillips R, Jeswani HK, Azapagic A, Apul D (2018) Are stormwater pollution impacts significant in life cycle assessment? A new methodology for quantifying embedded urban stormwater impacts. Sci Total Environ 636:115–123

Pourret O, Hursthouse A (2019) It's time to replace the term "heavy metals" with "potentially toxic elements" when reporting environmental research. Int J Environ Res Publ Health 16:4446

Pulster EL, Gracia A, Snyder SM, Romero IC, Carr B, Toro-Farmer G, Murawski SA (2020) Polycyclic aromatic hydrocarbon baselines in Gulf of Mexico fishes. In: Murawski S, Ainsworth CH, Gilbert S, Hollander DJ, Paris CB, Schlüter M, Wetzel DL (eds) Scenarios and responses to future deep oil spills. Springer, Cham, pp 253–271

Qin X, Wang F, Deng C, Wang F, Yu G (2016) Seasonal variation of atmospheric particulate mercury over the East China Sea, an outflow region of anthropogenic pollutants to the open Pacific Ocean. Atmos Pollut Res 7(5):876–883

Ramirez-Llodra E, Tyler PA, Baker MC, Bergstad OA, Clark MR, Escobar E, Levin LA, Menot L, Rowden AA, Smith CR, Van Dover CL (2011) Man and the last great wilderness: human impact on the deep sea. PLoS One 6(2011):e22588

Ramu K, Kajiwara N, Mochizuki H, Miyasaka H, Asante KA, Takahashi S, Ota S, Yeh HM, Nishida S, Tanabe S (2006) Occurrence of organochlorine pesticides, polychlorinated biphenyls, and polybrominated diphenyl ethers in deep-sea fishes from the Sulu Sea. Mar Pollut Bull 52:1827–1832

Rippy MA, Deletic A, Black J, Aryal R, Lampard JL, Tang JYM, McCarthy D, Kolotelo P, Sidhu J, Gernjak W (2017) Pesticide occurrence and spatio-temporal variability in urban run-off across Australia. Water Res 115:245–255

Romero-Romero S, Herrero L, Fernández M, Gómara B, Acuña JL (2017) Biomagnification of persistent organic pollutants in a deep-sea, temperate food web. Sci Total Environ 605:589–597
Roussiez V, Ludwig W, Monaco A, Probst J-L, Bouloubassi I, Buscail R, Saragoni G (2006) Sources and sinks of sediment-bound contaminants in the Gulf of lions (NW Mediterranean Sea): a multi-tracer approach. Cont Shelf Res 26:1843–1857
Rumampuk ND, Rumengan IF, Rompas RM, Undap SL, Boneka FB, Jensen KR, Lasut MT (2018) Tributyltin (TBT) contamination and impacts on imposex in Thalessa aculeata (Mollusca: neogastropoda: muricidae) in minahasa peninsula coastal waters, North Sulawesi, Indonesia. AACL Bioflux 11:184–193
Rusiecka D, Gledhill M, Milne A, Achterberg EP, Annett AL, Atkinson S, Birchill A, Karstensen J, Lohan M, Mariez C, Middag R, Rolison JM, Tanhua T, Ussher S, Connelly D (2018) Anthropogenic signatures of lead in the Northeast Atlantic. Geophys Res Lett 45(6):2734–2743. https://doi.org/10.1002/2017GL076825
Said TO, Farag RS, Younis AM, Shreadah MA (2006) Organotin species in fish and bivalve samples collected from the Egyptian Mediterranean coast of Alexandria. Egypt Bull Environ Contamin Toxicol 77:451–458. https://doi.org/10.1007/s00128-006-1086-8
Said TO, Shreadah MA, Abd El Ghani SA, Abd El Moniem MA (2010) Alkyltin and alkyllead compounds in coastal water of the Suez gulf, Egypt. Egypt J Aquat Res 36(1):33–42. https://www.scirp.org/reference/referencespapers?referenceid=1892306
Sarkar M, Islam JB, Akter S (2016) Pollution and ecological risk assessment for the environmentally impacted Turag River, Bangladesh. J Mater Environ Sci 7:2295–2304
Schlosser C et al (2016) Distribution and cycling of lead in the high and low latitudinal Atlantic Ocean. American Geophysical Union, Ocean Sciences Meeting 2016, abstract No. CT14B–0130
Schwarzenbach RP, Escher BI, Fenner K, Hofstetter TB, Johnson CA, von Gunten U, Wehrli B (2006) The challenge of micropollutants in aquatic systems. Science 313:1072–1077
Schwientek M, Guillet G, Rügner H, Kuch B, Grathwohl P (2016) A high-precision sampling scheme to assess persistence and transport characteristics of micropollutants in rivers. Sci Total Environ 540:444–454
Shreadah MA, Said TO, Younis AM, Farag RS (2006) Speciation of organotin compounds in sediments of semi-closed areas along the Mediterranean coast of Alexandria. Chem Ecol 22(5):395–404. https://doi.org/10.1080/02757540600917443
Shreadah MA, Said TO, Abd El Ghani SA, Abd El Moniem MA (2008) Alkyllead and alkyltin species in different fishes collected from the Suez gulf, Egypt. Egypt J Aquat Res 34(3):64–73. https://www.scimagojr.com/journalsearch.php?q=21100324710&tip=sid&clean=0
Shreadah MA, Said TO, Abd El Monem MI, Fathallah EMI, Mahmoud ME (2011) PAHs in sediments along the semi-closed areas of Alexandria. Egypt J Environ Protect 2(5):700–709. https://doi.org/10.4236/jep.2011.25063
Singh P, Sharma K, Hasija V, Sharma V, Sharma S, Raizada P, Singh M, Saini AK, Hosseini-Bandegharaei A, Thakur VK (2019) Systematic review on applicability of magnetic iron oxides–integrated photocatalysts for degradation of organic pollutants in water. Mater Today Chem 14:100186
Soerensen AL, Jacob DJ, Schartup A, Sunderland EM (2016) A mass budget for mercury and methylmercury in the Arctic Ocean. Glob Biogeochem Cycles 30(4):560–575
Spindola Vilela CL, Damasceno TL, Thomas T, Peixoto RS (2022) Global qualitative and quantitative distribution of micropollutants in the Deep Sea. Environ Pollut 307:119414. https://doi.org/10.1016/j.envpol.2022.119414
Stamm C, Räsänen K, Burdon J, Altermatt F, Jokela J, Joss A, Ackermann M, Eggen RIL (2016) Unravelling the impacts of micropollutants in aquatic ecosystems: interdisciplinary studies at the interface of large-scale ecology. Adv Ecol Res 55:183–223
Sumartias S (2020) The community responses over the Citarum river West Java. Eur J Mol Clin Med 7(1):3724–3732
Sun J, Yang Y, Fan S, Wang R, Huang Q (2024) The spatial variation in organotin contamination in different environmental media: a case of the lower Yangtze River in China. Water 16:3126

Takahashi S, Karri R, Tanabe S (2014) Contamination by persistent organic pollutants and related compounds in deep-sea ecosystems along frontal zones around Japan. In: Belkin IM (ed) The handbook of environmental chemistry. Springer, Berlin, pp 1–36. https://doi.org/10.1007/698_2013_252
Tanaka S, Kurasaki M, Morikawa M, Kamiya Y (2023) Design of materials and technologies for environmental remediation. Hdb Environ Chem 115:3–32. https://doi.org/10.1007/698_2021_819
Uc-Peraza RG, Delgado-Blas VH, Rendón-von Osten J, Castro ÍB, Proietti MC, Fillmann G (2022) Mexican paradise under threat: the impact of antifouling biocides along the Yucatán peninsula. J Hazard Mater 427:128162
UNEP (2011a) Hazardous substances in the Mediterranean: a spatial and temporal assessment. United Nations Environment Programme, Athens
UNEP (2011b) Analysis of trend monitoring activities and data for the MED POL phase III and IV (1999 2010). United Nations Environment Programme, Athens
UNEP (2012a) Initial integrat ed assessment of the Mediterranean Sea: fulfilling step 3 of the ecosystem approach process. United Nations Environment Programme, Athens
UNEP (2012b) State of the Mediterranean marine and coastal environment. United Nations Environment Programme, Athens
UNEP (2017) Chapter 17: shipping. In the first global integrated marine assessment: world ocean assessment I. Cambridge University Press, Cambridge
Usero J, Morillo J, Gracia I (2005) Heavy metal concentrations in molluscs from the Atlantic coast of southern Spain. Chemosphere 59:1175–1181
Van Cauwenberghe L, Vanreusel A, Mees J, Janssen CR (2013) Microplastic pollution in deep-sea sediments. Environ Pollut 182:495–499
Van den Broeck H, De Wolf H, Backeljau T, Blust R (2009) Comparative assessment of reproductive impairment in the gastropod mollusc Littorina littorea along the Belgian North Sea coast. Sci Total Environ 407:3063–3069. https://doi.org/10.1016/j.scitotenv.2008.12.050
Verhaegen Y, Parmentier K, Swevers L, Renders E, Rouge P, De Coen W, Cooreman K, Smagghe G (2011) The heterodimeric ecdysteroid receptor complex in the brown shrimp Crangon crangon: EcR and RXR isoform characteristics and sensitivity towards the marine pollutant tributyltin. Gen Comp Endocrinol 172:158–169
Verhaegen Y, Monteyne E, Neudecker T, Tulp I, Smagghe G, Cooreman K, Roose P, Parmentier K (2012) Organotins in North Sea brown shrimp (Crangon crangon L.) after implementation of the TBT ban. Chemosphere 86:979–984
Verslycke TA, Vethaak AD, Arijs K, Janssen CR (2005) Flame retardants, surfactants and organotins in sediment and mysid shrimp of the Scheldt estuary (The Netherlands). Environ Pollut 136:19–31
Vilela CLS, Bassin JP, Peixoto RS (2018) Water contamination by endocrine disruptors: impacts, microbiological aspects and trends for environmental protection environ. Pollution 235:546–559
Villarrubia-Gómez P, Cornell SE, Fabres J (2018) Marine plastic pollution as a planetary boundary threat—the drifting piece in the sustainability puzzle. Mar Pol 96:213–220. https://doi.org/10.1016/j.marpol.2017.11.035
Vogt EL, Model JFA, Vinagre AS (2018) Effect of organotins on crustaceans: update and perspective. Front Endocrinol 9:8
Waldock R, Rees HL, Matthiessen P, Pendle MA (1999) Surveys of the benthic infauna of the crouch estuary (UK) in relation to TBT contamination. J Mar Biol 79:225–232
Welsch H (2006) Environment and happiness: valuation of air pollution using life satisfaction data. Ecol Econ 58:801–813
White HK, Hsing P-Y, Cho W, Shank TM, Cordes EE, Quattrini AM, Camilli R, Demopoulos AWJ, German CR, Brooks JM, Roberts HH, Shedd W, Reddy CM, Fisher CR (2012) Impact of the deepwater horizon oil spill on a deep-water coral community in the Gulf of Mexico. Proc Natl Acad Sci USA 109:20303–20308. https://doi.org/10.1073/pnas.1118029109

WHO (2010) The world health report - health systems financing: the path to universal coverage. https://apps.who.int/iris/handle/10665/44371
Wirzinger G, Vogt C, Bachmann J, Hasenbank M, Liers, Stark C, Ziebart S, Oehlmann J (2007) Imposex of the netted whelk Nassarius reticulatus (Prosobranchia) in Brittany along a transect from a point source. Cah Biol Mar 48:85–94
Woodall LC, Sanchez-Vidal A, Canals M, Paterson GL, Coppock R, Sleight V, Calafat A, Rogers AD, Narayanaswamy BE, Thompson RC (2014) The deep sea is a major sink for microplastic debris. R Soc Open Sci 1:140317
Woodwell GM, Craig PP, Johnson HA (1971) DDT in the biosphere: where does it go? Science 174:1101–1107
Yadav P, Singh J, Srivastava DK, Mishra V (2021) Environmental pollution and sustainability. In: Singh P, Verma P, Perrotti D, Srivastava KK (eds) Environmental sustainability and economy. Elsevier, Amsterdam, pp 111–120
Younis AM (2020) Accumulation and rate of degradation of organotin compounds in coastal sediments along the Red Sea, Egypt. Egypt J Aquat Biol Fish 24(5):413–436. https://doi.org/10.21608/ejabf.2020.108918
Zandalinas SI, Fritschi FB, Mittler R (2021) Global warming, climate change, and environmental pollution: recipe for a multifactorial stress combination disaster. Trends Plant Sci 26:588–599
Zhakovskaya Z, Neevin IA, Metelkova L, Ryabchuk DV, Kukhareva G, Grigoriev AG, Egorova A, Krek AV (2022) Mobility of metal-organic pollutants in the emerging coastal-marine sediment of the Baltic Sea: the case example of organotin compounds in sediments of the Gulf of Finland. J Sea Res 190:102307
Zhang S, Li P, Li ZH (2021) Toxicity of organotin compounds and the ecological risk of organic tin with co-existing contaminants in aquatic organisms. Comp Biochem Physiol C Toxicol Pharmacol 246:109054
Zoller U (2006) Estuarine and coastal zone marine pollution by the nonionic alkylphenol ethoxylates endocrine disrupters: is there a potential ecotoxicological problem? Environ Int 32:269–272

Chapter 6
Control (Prevention) of OMCs, Future, and Sustainability

Abstract This chapter, which examines the entire issue, will facilitate progress in addressing the problem to the marine environment and humanity. OMCs exhibit distinctive reactivity profiles, ligand exchange kinetics, and redox potentials, rendering them suitable for various applications. The reader will gain an awareness of the fundamental needs, developments, and difficulties encountered in the aquatic environment from this chapter and investigate the consequences of OMCs on the marine environment, although joint efforts and preventive measures are being made to mitigate their detrimental implications on the marine ecosystem. Discussions of conclusions, trends, and viewpoints are included.

Keywords Control · Future · Sustainability · Prevention · OMCs

6.1 Preface

By integrating sustainability into their maintenance practices, industries are redefining success as global worries about the effects of our economy on the environment grow. As a result of this change, maintaining eco-friendly operations is now crucial. With their data-driven insights, computerized maintenance management systems (CMMS) assist businesses in maximizing energy use, cutting waste, and prolonging the life of their assets. CMMS software assists maintenance teams in centrally managing all of their maintenance tasks. It is part of planning, monitoring, measuring, and optimizing everything from work orders to inventory and reporting. In summary, industrial, asset-intensive firms' maintenance teams utilize CMMSs to manage information, ensure that tasks are completed, and assess the results of their labor. The connection between sustainability and maintenance not only supports international conservation objectives but also provides a practical and doable strategy for future-proofing industries (Goncalves et al. 2024).

T. O. Said, G. M. El Zokm, *Organometallic Compounds in the Marine Environment*, https://doi.org/10.1007/978-3-031-99189-9_6

6.1.1 Challenges

Even though OMC toxicology has advanced quickly in recent years, there are still several widely held “beliefs” that are prevalent in the OMC community despite their historical foundation. Existing opinions concerning (1) the possibilities of toxicity measurements, (2) the high toxicities of heavy-metal compounds, (3) the correlation between the structure of a metal compound and its toxicity, (4) the biological effect of direct/indirect contacts with OMCs, (5) dangers of metal nanoparticles, and (6) following a review of these topics, step-by-step descriptions of the fundamental ideas behind toxicity studies and known data are provided. This tutorial’s major objective is to show how a metal’s toxicity cannot be considered a constant attribute because it relies on several parameters, including its oxidation state, ligands, solubility, particle shape, environmental characteristics, and more. The biological impacts of such chemically labile species as metal complexes should not be presumed or taken for granted; precise measurements are necessary to draw valid findings (Egorova and Ananikov 2017). Many authors reported that physical, chemical, and biological processes in the environment mediate the environmental chemistry of OMCs (Haydee and Dalma 2017; Kucharski et al. 2022). Organic matter and biota aggregation for OMCs revealed strong coefficients of octanol-water [K_{ow}], organic carbon [K_{oc}] partitioning, and weak water solubility (Batista-Andrade et al. 2018). TBT-OH exhibits an acid-dissociation constant (pKa) of 6.3 for the TBT cation in saltwater. Additionally, cations of TBT and other OTs are drawn to and adsorbed onto suspended particles with net negatively charged surfaces, like organic matter and clay. The cationic form of TBT and TFT (Bu_3Sn^+ and Ph_3Sn^+) is stable at lower acidity levels; however, the two forms primarily exist as chlorides and hydroxides in an aqueous medium with a pH of 6–8 for ideal seawater (de Oliveira et al. 2010; Temerdashev et al. 2023). However, OMCs are also present as neutral molecules in aquatic environments. Their bioaccumulation, leading to a negative impact on aquatic systems, is relevant to cation form. Many extremely poisonous chemicals have been found in OMCs; this has been linked to the lipophilic characteristics of OMCs, which accelerate the rate of resorption. Most organometal[loid]s undergo metabolization to monovalent cations following resorption by losing an organic substituent. Due to their dual lipophilic and hydrophilic characteristics, these cations can easily pass through cell membranes and are transported through bodily fluids with efficiency. The main reason for this is that they may be readily and rapidly increased to very effective concentrations in target organs.

Because of the negative impacts of OMCs, efforts have been made to address the issue globally, and regulations have been put forth to safeguard the aquatic environment. Joint efforts by the research community, such as the European NORMAN (network of reference laboratories, research centers, and related organizations for monitoring of emerging environmental substances) or similar from other areas that link the research and policymaking worlds are highly valued. Numerous rules are in place regarding the harmful compounds found in OMCs, particularly the most prevalent compound [tributyltin (TBT)]. Tributyltin self-polishing copolymer paints [TBT-SPC paints], tributyltin oxide [TBTO, $C_{24}H_{54}OSn_2$], and tributyltin fluoride [TBTF, $C_{12}H_{27}SnF$] are predominant. According to the International Maritime Organization (IMO), a global convention that forbids the use of TBT-based paints beginning in 2003, and outright by 2008, is required. The EU's Water Framework Directive [WFD] and Marine Strategy Framework Directive [MSFD] aim to establish a good ecological and environmental state, respectively [WFD]. The primary community instrument for the management of point and diffuse discharges of hazardous materials in Europe is the present Water Framework Directive. Water Framework Directive 2000/60/EC was amended by Decision No. 2455/2001/EC on November 20, 2001, of the European Commission Parliament, defining 11 priority hazardous substances, including TBT compounds, subject to a ban on emissions, discharges, and losses into water (EU-SCOOP 2003, 2006; OECD 2007).

Furthermore, Regulation 782/2003 of 2003, on the prohibition of organotin, was accepted by the European Commission Parliament through Decision No. 415/2004/EC of March 2004, revising Regulation 2099/2002. The use of OTCs in antifouling paints is governed by the International Convention on the Control of Harmful Anti-fouling Systems on Ships, which began to take effect in 2001. Since 2008, the use of these compounds in ship paints has been completely prohibited (IMO MEPC 2008). Nonetheless, there are still current reports on BT levels and the bio-androgenic effects they linked to marine ecosystems, particularly in ports (Younis 2020; Beyer et al. 2022; Gomez et al. 2024). As a result, it's critical to keep an eye on changes in TBT concentrations and the ecological dangers they pose in the marine environment. We also need to develop TBT management plans and methods to lessen the effects of TBT on marine ecosystems. As a result, TBT has a half-life of 1.8–7 years in marine sediments. Furthermore, according to Lee et al., the half-life of TBT in anaerobic sediments is much greater, lasting more than 10 years. In addition to drawing attention to the environmental impact of OMCs, policies, awareness, and challenges are required for the accomplishment of the Sustainable Development Goals [SDGs], clean water and sanitation [G6], life below water [G14], and other goals. The issue of TBT-based antifouling has led to the adoption of regulations and guidelines by the IMO, which speaks for the global community, Anti-fouling Systems on Ships Convention [AFSC]. To achieve significant pollution reductions, the International Maritime Organization [IMO] is in charge of enforcing maritime pollution prevention accords like the London Convention, the AFSC, the International Convention for the Prevention of Pollution from Ships [MARPOL], and the London Protocol. It is essential to be willing to pay for government-provided amenities, social influence, clean water, and public awareness. Public awareness has a role in mediating relationships, which is a useful policy proposal for governments and other interested parties to strive toward achieving SDGs (IMO MEPC 2008; Kirval and Çalişkan 2022).

6.2 Joint Efforts by Different Networks and Worldwide Policymakers

The goal of waste management methods should be to choose the disposal method that has the least overall impact on harm to human health, disruption, and/or environmental damage, as well as the related social and economic consequences. The planning loop, the preliminary scientific assessment loop, and the monitoring and adaptation loop are its three phases or decision loops. To make decisions on integrated, environmentally friendly development planning, the plan acknowledges scientific and socioeconomic inputs as two concurrent, dynamic, and complementary processes. It highlights the independence and objectivity of scientific evaluations while simultaneously highlighting how deeply they impact judgments about socioeconomic viability. In the planning loop, an action is triggered by socioeconomic goals. The identification of current and prospective resources requires scientific evaluation. Choices must also be first considered as part of the process. Converting the specified environmental quality objectives (EQO) into the maximum permitted inputs is the fundamental task of the assessment process in the next step. Development activities and current and anticipated future pollutants are identified as part of the assessment process. Adopting water quality standards will entail identifying the most vulnerable group or target to be safeguarded and determining the contaminant's important pathway to it. To do this, water quality standards may be created using toxicity data for the target and the pollutant, as well as an appropriate application factor. The establishment of permissible inputs is made possible by the evaluation of the environmental capacity and the allocation of a portion of it for the specific project, using environmental data and end points based on water quality standards. There will always be several sources of uncertainty in such a process, necessitating experience-based approximations. The monitoring and adaptation loop illustrates this last and most crucial phase. Monitoring allows the project to become operational by testing if the environmental capacity is balanced. The project needs to be changed, or other technology would be needed, mainly in the waste and effluent treatment, if monitoring indicates that environmental capacity is exceeded. The project must be cancelled for environmental reasons if there isn't a technologically or economically viable substitute. In certain situations, permitted inputs may be raised if development and economic demands demand it, but only under strict supervision and with long-term monitoring data.

In the marine environment, several compounds are employed in ways that are purposefully dispersive. This does not mean that pollution won't happen, and if they are persistent and toxic enough, they could have unacceptably negative side effects. Recently, one such instance has emerged as a result of a certain type of antifouling paint. Fisheries scientists stated in 1980 that they thought the TBT used in antifouling paints used on pleasure vessels was the reason for shell deformation issues and recruitment failures in the Pacific oyster (*Crassostrea gigas*) business. Since then, additional research has verified that many other species were also negatively impacted, in addition to the Pacific oyster. Since the lowest concentration that has been shown to have an adverse effect is 0.06 μg/l (reduced rate of growth in Ostrea edulis spat), it is evident that the safe concentration for continuous exposure is below this value (possibly around 0.01 μg/l). However, it has not yet been possible to determine the concentration below which harmful effects on valued species are unlikely to occur. According to surveys, TBT concentrations in oyster-growing regions in an estuary where over 1700 boats may be moored during the summer months might surpass 0.20 μg L^{-1}, which is far higher than the 0.06 μg L^{-1} value at which adverse effects emerge. TBT can be bound in a copolymer composition to regulate the rates of release from antifouling paints. This is an improvement compared to the earlier "free-association" paints, which release a significant percentage of their TBT content during the first 4 weeks of immersion. The slower controlled-release copolymer kind of paint has already seen a significant shift. To determine the TBT leach rates under various temperature ranges, tests have been carried out using both types of paint. Under summer circumstances, the initial release rate was around 40 μg/cm^2 day, but after about 2 months, it had decreased to roughly a tenth of this amount, according to experiments using the free-association type. Copolymer paint leach rates typically vary from 6.00 to 11.00 μg cm^{-2} day^{-1} under similar conditions. By making the very basic assumption that all moored boats are painted with copolymer-based paints or free-association type paints, without accounting for tidal flushing or other removal processes, it is possible to estimate the maximum leach rates into an estuary. However, since the maximum rate of exchange with the open sea is known to be around 5% per day, it is easy to account for tidal flushing of the estuary in question. It was assumed, conservatively, that additional elimination methods could be disregarded. It can be calculated that if all of the 1700 boats moored in the estuary were painted with free-association paint, the maximum amount of TBT present in the estuary would be around 27 kg, falling to about 7 kg after 2 months. This is based on a volume of 50×10^6 m^3 and the assumption that the mean hull length of the boats is 8 m. Estuarine concentrations would fall between 0.500 and 0.150 μg l^{-1} as a result. Similarly, the concentration in the water would increase to around 0.2 μg/l (the maximum amount recorded in the water away from the immediate area of the boats) if all the boats were painted with a copolymer paint that had the lowest leach rate (6.00 μg cm^{-2} day^{-1}). Given that both values are far higher than what would be deemed acceptable (<0.020 μg L^{-1}), the environmental capacity would be exceeded. Less than 0.2 μg/cm^2 may be required for the leach rate to completely eradicate the risk of contamination in the estuary; even lower leach rates may be required in places with more restricted water volumes and exchange, such as marinas or lagoons, or if the number of boats is increased. Depending on the testing technique, leach rate data vary; some are lower than those used here, but even these lower rates produce concentrations in estuaries that are far higher than those that are probably deemed safe. By limiting the use of these paints on recreational vessels, several national authorities have already taken action to safeguard their oyster resources. In order to restrict input (to what is considered safe), another nation has followed the environmental capacity method. They have suggested that the leach rate of paints should not be greater than 0.100 μg cm^{-2} day^{-1}. In one nation, control measures were implemented. Both the quality of the oysters in spatial recruitment and the decreased levels of TBT in the water have shown notable improvements since then. This example illustrates the challenge of managing a seasonal and diffuse input of a highly toxic compound, for which the absence of a routinely applicable, appropriately sensitive analytical method to detect the extremely low environmental concentrations known to be harmful has hampered the derivation of a safe level. The suggested preliminary water quality limits of 0.01 μg/l reflect the extremely low environmental capacity of this herbicide, which is employed intentionally to harm marine life. It is also known that TBT is toxic to humans, and the absence of toxicological evidence to support an acceptable daily intake (ADI) will make it difficult to fully evaluate the hazards to this target and potential exposure routes. Therefore, it is not possible to do a complete critical pathway analysis at this time (IMO/FAO/UNESCO/WMO/WHO/IAEA/UN/UNEP 1986). Mercury has been seen as a material whose discharge should be strictly regulated ever since the Minamata Bay event, in which several individuals were poisoned by mercury in fish and shellfish that had been polluted by an industrial discharge containing mercury. Fish tissue concentrations in many sections of a European nation were much higher than those in other locations, according to a 1971 investigation of the mercury content of fish collected at different sites. Two sources of mercury were found by these investigations: sewage sludge from the works that was transported by ship and deposited in the impacted region and treated sewage effluent that was released into the outer reaches of the surrounding estuary. A factory was the primary source of mercury, and its wastewater was dumped into the public sewer. Quick action was made to cut down on this mercury intake, but it was also necessary to determine the highest amount that could be released without risk. A thorough review of the literature revealed that the mercury amounts that were then detected in the impacted area's waters (a maximum of 50 ng/l) were far below levels that, even after prolonged exposure to the tainted waters, might endanger marine life. Therefore, even though fish in particular had abnormally high amounts of mercury in their muscular tissue, they were not in danger. Nonetheless, the edible tissue of a typical assortment of fish that the local populace is expected to eat had mercury levels of around 0.45 mg day^{-1}. Out of a total estimated intake of 25.4 kg day^{-1}, the primary sources of mercury in the affected region (sewage effluents and deposited sludges) contributed 24.8 kg day^{-1}. As a result, fish had an average concentration of 0.45 mg/kg. A weekly meal of between 650 and 700 g of fish may cause the provisional tolerated weekly intake (PTWI) of mercury to surpass the FAO/WHO-recommended limit of 0.3 mg. As a result, while fish were not in danger, those who consumed more fish than the nation's average of less than 120 g/week may be at risk. The maximum amount of mercury that may be released into the region each day is 10 kg. Because of the steps implemented at the industrial site to limit mercury emission, this aim was readily met, and inputs have been gradually lowered to the current level of around 3.5 kg/day. Following critical path analysis, the most sensitive target was identified, and the environmental capacity was determined. It was determined that BPMA would more than meet the needed decrease in wastewater, which was essential.

Man's utilization of metalliferous ores to extract the metal in question is one of the first examples of "industrial" waste. At a location on the shore of a sizable estuary, a sizable smelting plant handles scrap and nonferrous metal ores. Along with the metals that were recovered—cadmium, copper, zinc, nickel, and lead—significant levels of arsenic were also found. Along with the process of water, decisions had to be made on the amounts and concentrations of the different elements that may be safely released into the estuary. Estuarine water quality standards of 5, 5, 40, 30, and 25 μg/l have been established based on an evaluation of the literature on the acute toxicity and possible long-term effects of cadmium, copper, zinc, nickel, and lead. These standards are thought to be sufficient to protect marine life outside of the immediate mixing zone. They were implemented without additional evaluation and suggested stringent limitations on the discharge. The estuary at the site under consideration is sizable and has good flushing. There is significant mixing, and the tidal effluents are extremely powerful. It was believed that the estuary's typically high suspended solids load would effectively absorb the arsenic and make it less accessible to marine life. According to recent evaluations of the potential risks of arsenic to humans, even a small amount of inorganic arsenic may be hazardous, and fish flesh with 1.00 mg kg^{-1} wet weight of the metal may be dangerous. As a result, recent fish samples have been examined for both total and inorganic arsenic. It has been determined that the majority of arsenic is, as anticipated, in an organic form and that inorganic arsenic concentrations seldom ever surpass 1.00 mg kg^{-1} wet weight. Arsenic from abandoned mine workings and smelter spoil piles, which have existed in some places for hundreds of years, is a problem in many coastal locations. Based on experience at these locations, it appears that arsenic releases into the marine environment are not always linked to harm to marine life or arsenic buildup by local species. Additionally, it is well recognized in the area that the majority of arsenic that marine species acquire is in an organic form, which is comparatively inaccessible to humans and does not pose a risk. It was determined that release to the estuary would be the preferred and safest choice for handling the arsenic due to the risks associated with the other two options. However, because some fish species may accumulate arsenic, a monitoring mechanism was established to make sure that, if this occurs, corrective action may be taken right away and the danger to humans as fish consumers is minimized.

Sensing technology is becoming more and more popular, and it is even being proposed as a part of personalized healthcare management. This is because of the potential applications that bioengineering-supported organometallic compounds have. OMCs have abundant biological activity, numerous industrial and material science applications, and unique biological targetability that makes them potentially useful for drug development, diagnosis, and treatment. OMCs are also becoming more and more of a viable option for sensor technology advancement because they are designed to overcome some of the drawbacks of their conventional counterparts (Mishra et al. 2023). These compounds have distinctive characteristics that render them exceptionally responsive to alterations in their surroundings, hence enabling their use as sensors for the detection of diverse chemicals or circumstances. In addition, the adaptability of OMCs allows for their incorporation into various sensor platforms, rendering them appropriate for a diverse array of applications such as environmental surveillance, medical analysis, and industrial assurance of quality. This review presents an overview of the recent progress made in the field of OMCs' design and production, specifically focusing on their potential uses as sensors. In addition, the structural modifications, functionalization procedures, integration of microfluidics, and the consequent effects on the sensing capacities of the materials are recalled. These bio-based methods will support sustainability goals and increase the effectiveness, affordability, and applicability of sensing. OMCs are emerging as promising candidates for sensor development owing to their unique properties. Their utilization as sensors has significantly improved the identification and monitoring of hazardous gases in industrial manufacturing (Bastani et al. 2016), facilitated atmospheric environmental analysis (Lieberzeit and Dickert 2007), and contributed to the prevention of fire and explosion incidents (Ghosh and Dhar 2019). The importance of their applications is widely acknowledged in the current literature review, as shown by the large number of book chapters, reviews, and research articles that have been published recently. However, selectivity, stability, and other aspects related to the development of sensitive materials are currently lacking. As a result, research and development into new high-performance sensors have great practical value and utility (Vigneshvar et al. 2016). Chemical sensors typically detect chemicals and then produce chemical signals as a result (Fracchiolla et al. 2013). Photochemical sensors and electrochemical sensors are the two main categories into which chemical sensors can be divided (Vigneshvar et al. 2016). Chemical sensors find widespread application in sensing applications, including industrial automation (Turner 2013), meteorological observation, process analysis, resource detection, telemetry, fresh food preservation, production (Wang et al. 2014), energy conservation (Abe et al. 2014), security alarms, environmental pollution monitoring (Wang et al. 2013), and remote diagnosis (Harris et al. 2013). The human gustatory and olfactory systems share many similarities with the underlying mechanisms of chemical sensors (Mello et al. 2013). To enhance patient monitoring and medical diagnostics, researchers have recently demonstrated a strong interest in employing organometallic compounds as biosensors in the healthcare industry. The term "biosensor" refers to an extremely sophisticated analytical apparatus that integrates a biological sensing component. This type of equipment can be used for a wide range of purposes, such as drug development, biomedicine, food processing and safety, environmental monitoring, defense, and security. The original biosensor was created in 1962 by Clark and Lyons to measure the amount of glucose present in biological samples. This biosensor used an electrochemical detection technique to identify hydrogen peroxide or oxygen using an immobilized glucose oxidase electrode (Jamal et al. 2012).

The use of OMCs, which combine metal and organic active moieties, in batteries has demonstrated great promise. OMCs' varied structure, abundant porosity, and distinct charge centers allow them to work in batteries. First, we categorize OMCs in this review according to their metal-organic frameworks, phthalocyanine, ferrocene, etc. Next, we primarily concentrate on each type of OMC's recent advancement in rechargeable batteries. In a range of battery systems, including Li, Na, Zn,

Li-S, Li-O_2, and redox flow batteries, they can function as electrode or electrolyte materials. The structures and characteristics of OMCs have a direct bearing on the performance improvements in these batteries. Comprehensive evaluations of the relationships between structure and performance are given (Wang et al. 2021).

6.3 Clinical Use of OMCs

There is a wide variety of OMCs or complexes in the natural world. Their speciation—the entire molecular identity of the entity in question—determines their importance, which typically translates to their potential for transport or toxic impact. Actual toxicity, the ability for transport within organisms and the geosphere, stability in the presence of light, heat, and water, and breakdown within organisms are all determined by speciation. This can be demonstrated by pointing out that arsenic oxides and their essential nontoxicity can both be found naturally in the environment, as can arsenobetaine. The varying toxicities of the different alkyltin species serve as evidence of the significance of speciation to toxicity (Jenkins et al. 2007). A flexible platform for creating therapeutics, ranging from antimicrobials to anticancer agents, is provided by OMCs. Titanocene dichloride development in clinical trials is one early success story. This unequivocally showed that organometallic compounds have potential for biological applications, even though it was unsuccessful. As the number of structural motifs employed increases, we are gaining a better understanding of how each component affects both the anticancer activity and the qualities needed to have the potential to be developed further, such as stability and compatibility with aqueous environments. Biological activity has been tested for almost every "classic" organometallic compound class due to the increasing number of researchers working in the field of bioorganometallic chemistry (Hanif and Hartinger 2018).

To mention a few, these include arenes, cyclopentadienes, carbonyls, carbenes, and metallocenes. They have all been shown to be beneficial in one way or another, whether it be for introducing stability, ligand release, redox qualities, or prompt ligand exchange. Unfortunately, no precise structure-activity correlation has been found between the various classes of anticancer organometallic compounds; this is because structural variations can frequently have a significant effect on biological properties, making it challenging to compare compounds within the same class. However, the organometallic scaffold's easy access to structural modifications enables fine-tuning of the biological properties to produce different modes of action from the traditional platinum-based anticancer metallodrugs.

The body's natural response to tissue damage, mostly brought on by chemical and physical factors, is inflammation (Kundu and Surh 2008). Long-term genotoxic processes or inadequate tissue repair may result in the development of cancer (Elinav et al. 2013). For inflammatory bowel diseases (IBD) to develop, intestinal inflammation mediated by the immune system is the most critical factor (Gentschew and Ferguson 2012). Lipopolysaccharide (LPS), tumor necrosis factor-alpha

(TNF-α), interleukin-1 (IL-1), or interferon-gamma (IFN-γ) can all stimulate it. Activated macrophages contribute to the increased production of cytokines and other inflammatory mediators following the start of the inflammatory process (Zhang et al. 2017, 2024). As a result, substances released by macrophages, like nitric oxide (NO), aid in escalating the inflammatory response (Lawrence et al. 2002). The transcription factors nuclear factor-κB (NF-κB) and signal transducer and activator of transcription 3 (STAT3) are the primary intracellular pathways linked to the emergence of inflammation (Elinav et al. 2013). Finding molecules with distinct anticancer mechanisms of action (MOA) from platinum (II) drugs is desperately needed, especially for patients who relapse after receiving a platinum-containing chemotherapy regimen (which accounts for about 50% of all anticancer chemotherapy treatments) during their initial treatment (Dasari and Tchounwou 2014). The antiproliferative activity of the 16-e complex [(*p*-cymene) Ru(maleonitriledithiolate)], toward ovarian (A2780 and A2780cisR), colon (HCT116 $p53^{+/+}$ and HCT116 $p53^{-/-}$), and lung (H460) cancer cells, is reported. Soldevila-Barreda et al. (2020) made efforts to confirm/refute the anticancer potential of such electron-deficient OMCs. The effectiveness of platinum antitumor chemotherapeutic medications has been hampered by issues with toxicity, drug resistance, and broad spectrum of action; yet, patients with solid malignancies treated with platinum therapies, most notably "cisplatin," have very good survival rates. As a result, a lot of chemical entities based on non-platinum metals are receiving attention and have progressed to preclinical testing and clinical trials. However, at a later stage, they are unable to be classified as medications, which causes significant setbacks for pharmaceutical research and development. Consequently, there is a drive to develop innovative metal-based cancer chemotherapeutics that demonstrate an alternative molecular mechanism of action for cell death (Arjmand et al. 2014). Among the non-platinum metal-based medications, organotin compounds have demonstrated their worth in effective control of toxicity issues and precisely targeted drug uptake exclusively by the malignant cells, leaving the healthy cells unaffected (apoptosis). They examined the milestones for the future predictions of organotin compounds in the drug industry and reflected on the progress made by these compounds over the last 10 years as antitumor chemotherapeutic agents (it was noted that over 50% of organotin compounds exhibit high cytotoxic activity but, surprisingly, have not entered clinical trials).

OMCs have attracted interest recently in the development of metal chemotherapeutics due to their remarkable stability and unique features relative to traditional coordination complexes. Their complexes possess unique reactivity profiles, ligand exchange kinetics, and redox potentials, making them ideal for controlling drug release and enhancing cellular absorption. The antiviral, anticancer, antibacterial, antifungal, and antiparasitic properties of metal complexes have been the subject of several investigations. Numerous complexes possessing intriguing biological characteristics have been created. Notably, certain metal complexes are used to cure many illnesses. Nitin et al. (2024) have offered an overview of the OMCs utilized in medicine, especially those with antibacterial and anticancer properties, complexes used to treat neurological diseases, and challenges faced in this particular discipline. The primary objective of their review is to provide readers with a summary of the significant progress made in each of these numerous medical professions regarding organometallics. They represented the overall function of OMCs in the realm of medicine and therapy. OMCs, which are composed of a metal and an organic component, are fascinating compounds in and of themselves. In 1908, Paul Ehrlich was awarded the Nobel Prize for his use of organometallics in medicine. Salvarsan (the first successful treatment for the deadly infectious diseases syphilis, relapsing fever, and African trypanosomiasis) was developed with this antibiotic medication at the start of the 1910s. Numerous studies have been conducted in the past several years, leading to the discovery of numerous organometallics that are now employed in various medical therapies. The fact that the cells or organisms get resistant to these compounds is the greatest enduring issue with them. The first contemporary antibacterial agent was this organoarsenic molecule. The usage of these complexes began in the treatment of disorders, which further emerged as a turning point in the therapeutic arena. There are various forms of organometallic complexes based on the type of link between the atoms, the metal that makes up the complex, or both.

The rising burden of antimicrobial resistance (AMR) is one of the most concerning risks to public health worldwide. With limited economic incentives and a declining supply of new medications in clinical pipelines, more inventive approaches to novel drug design and development are sorely required. Due to their capacity to eradicate infections through metal-specific mechanisms of action, metal-based compounds are quickly replacing organic medications in medical practice. Waters et al. (2023) examined developments in metal-based antibacterial agents, such as catalytic metallodrugs, metal ions, and complexes. Their review's conclusion presents a viewpoint on the sensible development of metal-based antibiotics and how they might take advantage of their special qualities to combat AMR. Metal complexes can bind to biomolecules covalently, as in the case of the DNA adduct formation with cisplatin. They covered the most recent research in antibiotic development, including various metal complexes and their various modes of action and targeting qualities, preferably conducted within the last 5 years. Other examples of metal-specific modes of action include redox-active or catalytically active drugs (i.e., metals that produce reactive oxygen species or ROS) and photoactivatable metal complexes (i.e., for photodynamic therapy (PDT)) (Boros et al. 2020).

By using interfacial polymerization systems, which involve the interaction of difunctional Lewis bases with organotin halides, organotin polymers can be produced with ease. Numerous polymeric products containing organotin have been produced, including the first ones that are soluble in water. Many different cancer cell lines, including those linked to ovarian, colon, lung, prostate, pancreatic, and breast cancer, are inhibited by the organotin polymers, among their many other biological functions (Carraher Jr and Roner 2014). Certain viruses also suppress a range of other viruses, such as those that cause smallpox and herpes infections. They described the synthesis, structural characterization, and biological characterization of these materials.

6.4 Conclusion, Trends, and Perspectives

The quantity of different pollutants in the environment has grown along with the industrial, agricultural, and pharmaceutical sectors' rapid development, posing a serious threat to ecosystems and human health (Yang et al. 2018, 2023; Liu et al. 2021; Souza et al. 2021). The wastewater produced during the refinement of minerals and metals contains a large amount of inorganic ions, which are typically persistent pollutants, some of which are even thought to be carcinogenic (Zhang et al. 2017; Chen et al. 2021). An overabundance of these contaminants can upset and harm a person's body. Pesticide and antibiotic residues in the environment have long drawn a lot of attention in pollution monitoring because they are organic pollutants (Qadeer et al. 2023). Using pesticides on crops during agricultural production is advantageous (Wei et al. 2023). On the other hand, the balance of the aquatic ecosystem could be harmed by the overuse of pesticides that seep into surface water bodies from agricultural land (Li et al. 2023a). Antibiotics are also common examples of toxic organic pollutants. High environmental concentrations of antibiotic residues have the potential to foster the growth of super-bacteria, with potential sources including pharmaceutical industry waste and sewage from farms that raise poultry or cattle (Anthony et al. 2018; Broom et al. 2023; Li et al. 2023b; Yadav et al. 2023a). Currently, atomic absorption spectrometry (Yadav et al. 2023b), gas chromatography (Wang et al. 2020), high-performance liquid chromatography (Ocampo et al. 2023), ion mobility spectrometry (Amini et al. 2021), and liquid chromatography-tandem mass spectrometry (Pallavi et al. 2023) are the most common methods used to detect antibiotic residues. These methods are time-consuming and have high operating costs, space requirements, and expensive instruments. Therefore, it is imperative to create a new fluorescent substance sensing-based detection technique that is quick to respond, inexpensive, and simple to use for a variety of pollutants (Bin et al. 2023). Coordination bonds allow a central metal ion and one or more ligands to combine to form molecules known as metal complexes (Kumar et al. 2017, 2023). The kind and structure of the ligand greatly affect its chemical and physical characteristics (Pathan and Patel 2023). Because of the easy combination and flexible coordination pattern of heterocyclic nitrogen compounds to form complexes with metal ions, organic ligands and ligands containing carboxylic groups have been extensively studied in coordination chemistry (Retnam et al. 2023). OMCs integrate the benefits of the exceptional optical qualities of both organic and inorganic fluorescent materials Their structures can be created, altered, and modified by using metal centers and organic ligands and transformed to become a new kind that can recognize particular ions and molecules of fluorescent material with multiple uses (Li et al. 2019; Ziolkowska et al. 2023). Metal complexes are frequently employed in the detection of different organic and inorganic contaminants (Li et al. 2023c). Thus far, metal complexes have demonstrated remarkable fluorescence properties and intriguing applications in environmental pollution detection (Das et al. 2022; Fang et al. 2023). In our opinion, OMCs will yield superior outcomes in ecological detection.

The use of OMCs in organic chemistry is a cornerstone of the modern synthetic technique for the synthesis and activation of new bonds in a molecule. The ability of transition metal complexes to catalyze organic reactions is one of the most powerful tools now available for developing novel synthetic pathways, even if stoichiometric transformations constitute the core of their applications. Zhu et al. (2024) involved the synthesis of a fluorescent complex, which is $[Zn(NTD)_2(DTP)_2(H_2O)_2]\cdot(H_2O)_{0.8}$ (Complex Zn), (H_2NTD = 1,4-naphthalene dicarboxylic acid and DTP = 3,5-di(1,2,4-triazol-1-yl)pyridine). Thermogravimetric, elemental, infrared spectroscopy, powder X-ray diffraction, and fluorescence analyses were used to characterize the fluorescent complex. Complex Zn demonstrated exceptional fluorescence quenching efficiency in the fluorescence sensing tests when tested in aqueous media against Fe^{3+}, MnO^{4-}, $Cr_2O_7^{2-}$, nitrofurantoin, and imidacloprid. According to a mechanism study, the inner filter effect and the fluorescence resonance energy transfer effect in the fluorescent sensing process are what cause the fluorescence to quench, which is caused by the quenchers moving toward the sensor. The complex's ability to quench its fluorescence in the presence of ions, pesticides, and antibiotics is largely dependent on the inner filter effect (IFE) and fluorescence resonance energy transfer (FRET).

References

Abe K, Yoshida W, Ikebukuro K (2014) Electrochemical biosensors using aptamers for theranostics. Adv Biochem Eng Biotechnol 140:183–202

Amini S, Ebrahimzadeh H, Seidi S, Jalilian N (2021) Preparation of polyacrylonitrile/Ni-MOF electrospun nanofiber as an efficient fiber coating material for headspace solid-phase microextraction of diazinon and chlorpyrifos followed by CD-IMS analysis. Food Chem 350:129242

Anthony AA, Adekunle CF, Thor AS (2018) Residual antibiotics, antibiotic resistant superbugs and antibiotic resistance genes in surface water catchments: public health impact. Phys Chem Earth 105:177–183

Arjmand F, Parveen S, Tabassum S, Pettinari C (2014) Organo-tin antitumor compounds: their present status in drug development and future perspectives. Inorg Chim Acta 423:26–37. https://doi.org/10.1016/j.ica.2014.07.066

Bastani K, Rao PK, Kong Z (2016) An online sparse estimation-based classification approach for real-time monitoring in advanced manufacturing processes from heterogeneous sensor data. IIE Trans 48(7):579–598

Batista-Andrade JA, Caldas SS, Batista RM (2018) From TBT to booster biocides: levels and impacts of antifouling along coastal areas of Panama. Environ Pollut 234:243–252

Beyer J, Song Y, Tollefsen KE, Berge JA, Tveiten L, Helland A, Øxnevad S, Schøyen M (2022) The ecotoxicology of marine tributyltin (TBT) hotspots: a review. Mar Environ Res 179:105689. https://doi.org/10.1016/j.marenvres

Bin DN, Kurdi A, Alshihri S, Tabbakh T (2023) Organic heterocyclic-based colorimetric and fluorimetric chemosensors for the detection of different analytes: a review (from 2015 to 2022). Mater Today Chem 27:101347

Boros E, Dyson PJ, Gasser G (2020) Classification of metal-based drugs according to their mechanisms of action. Chem 6:41–60

Broom A, Peterie M, Ridge D, Lafferty L, Kenny K, Broom J, Kelly HA, Treloar C, Applegate T (2023) Sex, drugs and superbugs: the rise of drug-resistant STIs. SSM Qual Res Health 4:100310

Carraher Jr CE, Roner MR (2014) Organotin polymers as anticancer and antiviral agents. J Organomet Chem 751:67–82. https://doi.org/10.1016/j.jorganchem.2013.05.033

Chen YG, He XLS, Huang JH, Luo R, Ge HZ, Wołowicz A, Wawrzkiewicz M, Gładysz-Płaska A, Li B, Yu QX (2021) Impacts of heavy metals and medicinal crops on ecological systems, environmental pollution, cultivation, and production processes in China. Ecotoxicol Environ Saf 219:112336

Crabtree RH (2009) The organometallic chemistry of the transition metals. Wiley, Hoboken

Das R, Pal R, Bej S, Mondal M, Kundu K, Banerjee P (2022) Recent progress in D optical nanoprobes for applications in the sensing of (bio)analytes with the prospect of global health monitoring and detailed mechanistic insights. Adv Mater 3:4421–4459

Dasari S, Tchounwou PB (2014) Eur J Pharmacol 2014(740):364–378

de Oliveira CR, dos Santos DM, Madureira LAD, de Marchi MRR (2010) Speciation of butyltin derivatives in surface sediments of three southern Brazilian harbors. J Hazard Mater 181(1–3):851–856. https://doi.org/10.1016/j.jhazmat.2010.05.091

Egorova KS, Ananikov VP (2017) Toxicity of metal compounds: knowledge and myths. Organometallics 36(21):4071–4090

Elinav E, Nowarski R, Thaiss CA, Hu B, Jin C, Flavell RA (2013) Inflammation-induced cancer: crosstalk between tumors, immune cells, and microorganisms. Nat Rev Cancer 13:759–771. https://doi.org/10.1038/nrc3611

EU-SCOOP (2003) Assessment of the dietary exposure to organotin compounds of the population of the EU member states, report on tasks for scientific cooperation. Directorate General Health and Consumer Protection. http://ec.europa.eu/food/food/chemicalsafety/contaminants/scoop_3

EU-SCOOP (2006) Revised assessment of the risks to health and the environment associated with the use of the four organotin compounds TBT, DBT, DOT and TPT. Directorate General Health and Consumer Protection. http://ec.europa.eu/health/ph_risk_committees/04_scher/docs/scher_o_047.pdf

Fang WQ, Zhang L, Feng HR, Meng JX, Zhang ZP, Liu ZB (2023) Research progress of fluorescent-substance@MOFs. Microchem J 185:108265

Fracchiolla NS, Artuso S, Cortelezzi A (2013) Biosensors in clinical practice: focus on oncohematology. Sensors 13(5):6423–6447

Gentschew L, Ferguson LR (2012) Role of nutrition and microbiota in susceptibility to inflammatory bowel diseases. Mol Nutr Food Res 56:524–535. https://doi.org/10.1002/mnfr.201100630

Ghosh P, Dhar PK (2019) GSM based low-cost gas leakage, explosion and fire alert system with advanced security. In: Proceedings of the international conference on electrical, computer and communication engineering (ECCE). IEEE, Piscataway, pp 1–5

Gomez NA, Sturla LJ, Ferrando A, Garrido M, Domini CE (2024) Update on the status of the contamination by organotin compounds in sediment of Nuevo gulf, Argentina. Insights from field and experimental studies. Mar Pollut Bull 200:116087. https://doi.org/10.1016/j.marpolbul.2024.116087

Goncalves T, Wallace E, Afara J (2024) Building a sustainable future with the help of maintenance and CMMS. https://fiixsoftware.com/blog/cmms-and-sustainability

Hanif M, Hartinger CG (2018) Organometallics in cancer treatment—non-conventional structures and modes of action. In: Reference module in chemistry, molecular sciences and chemical engineering. Elsevier, Amsterdam. https://doi.org/10.1016/B978-0-12-409547-2.14249-0

Harris JM, Reyes C, Lopez G (2013) Common causes of glucose oxidase instability in in vivo biosensing: a brief review. J Diabetes Sci Technol 7(4):1030–1038

Haydee KM, Dalma KE (2017) Concerning organometallic compounds in environment: occurrence, fate, and impact. In: Recent progress in organometallic chemistry. IntechOpen, London. https://doi.org/10.5772/67755

IMO MEPC (2008) IMO meeting approves revised regulations on ship emissions. IMO, London

IMO/FAO/UNESCO/WMO/WHO/IAEA/UN/UNEP (1986) Environmental capacity an approach to marine pollution prevention. Food and Agriculture Organization, Rome

Jamal A, Rahman MM, Khan SB, Faisal M, Akhtar K, Rub MA, Al-Youbi AO (2012) Cobalt-doped antimony oxide nano-particles based chemical sensor and photo-catalyst for environmental pollutants. Appl Surf Sci 261:52–58

Jenkins RO, Craig PJ, Francesconi KA, Harrington CF (2007) Environmental and biological aspects of organometallic compounds. In: Comprehensive organometallic chemistry III. From fundamentals to applications, vol 12. Elsevier, Amsterdam, pp 603–661. https://doi.org/10.1016/B0-08-045047-4/00178-3

Kirval L, Çalişkan UK (2022) Influence of the European Union [EU] on the international maritime organization [IMO] about the market-based measures on emissions. Int J Environ Geoinf. https://doi.org/10.30897/ijegeo.1047467

Kucharski D, Stasiuk R, Drzewicz P, Skowronek A, Strzelecka A, Mianowicz K, Giebułtowicz J (2022) Fit-for-purpose assessment of QuEChERS LC-MS/MS methods for environmental monitoring of organotin compounds in bottom sediments of the Odra River estuary. Molecules 27:4847. https://doi.org/10.3390/molecules27154847

Kumar G, Guda R, Husain A, Bodapati R, Das SK (2017) A functional Zn(II) metallacycle formed from an n-heterocyclic carbene precursor: a molecular sensor for selective recognition of Fe3+ and IO4− ions. Inorg Chem 56:5017–5025

Kumar A, Avula B, Yadav S, Singh Y, Eraganaboyina S, Zyryanov GV, Nemallapudi BR, Singh AK, Bhimireddi R (2023) Synthesis, properties, environmental stability and practical application of organometallic compounds: a comprehensive review. Inorg Chem Commun 158(2):111567. https://doi.org/10.1016/j.inoche.2023.111567

Kundu J, Surh Y (2008) Inflammation: gearing the journey to cancer. Mutat Res Mutat Res 659:15–30. https://doi.org/10.1016/j.mrrev

Lawrence T, Willoughby DA, Gilroy DW (2002) Anti-inflammatory lipid mediators and insights into the resolution of inflammation. Nat Rev Immunol 2:787–795. https://doi.org/10.1038/nri915

Lee S-H, Chen Y-S, Chen C-F, Albarico FPJB, Lim YC, Wang M-H, Chen C-W, Dong C-D (2022) Butyltin contamination in fishing port sediments after the ban of tributyltin antifouling paint: a case of Qianzhen fishing port in Taiwan. Water 14:813. https://doi.org/10.3390/w14050813

Li B, Xu ZQ, Xu YB, Yong GP (2019) Effects of substituent groups on the crystal structures and luminescence properties of zero-/two-dimensional Zn(II) complexes. Inorg Chem Commun 102:57–60

Li WT, Xin SH, Deng WJ, Wang BB, Liu XX, Yuan Y, Wang SL (2023a) Occurrence, spatiotemporal distribution patterns, partitioning and risk assessments of multiple pesticide residues in typical estuarine water environments in eastern China. Water Res 245:120570

Li S, Ondon BS, Ho SH, Li FX (2023b) Emerging soil contamination of antibiotics resistance bacteria (ARB) carrying genes (Args): new challenges for soil remediation and conservation. Environ Res 219:115132

Li S, Sun JY, Liu GC, Zhang S, Zhang Z, Wang XL (2023c) A new keggin-type polyoxometallate-based bifunctional catalyst for trace detection and pH-universal photodegradation of phenol. Chin Chem Lett 21:109148

Lieberzeit PA, Dickert FL (2007) Sensor technology and its application in environmental analysis. Anal Bioanal Chem 387:237–247

Liu Y, Wang P, Gojenko B, Yu JJ, Wei LZ, Luo DG, Xiao TF (2021) A review of water pollution arising from agriculture and mining activities in Central Asia: facts, causes and effects. Environ Pollut 291:118209

Mello L, Kisner A, Goulart M, Kubota L (2013) Biosensors for antioxidant evaluation in biological systems. Comb Chem High Throughput Screen 16(2):109–120

Mishra DP, Acharya B, Tripathy S, Barik B, Sahu PK (2023) An overview of the biosensing potential of organometallic compounds. Chem Phys Impact 7:100326. https://doi.org/10.1016/j.chphi.2023.100326

Nitin V, Anand T, Singh CK, Sodhi KK (2024) Role of organometallic complexes in targeted therapies of different diseases: infectious diseases, cancer and neurodegenerative diseases. J Organomet Chem 1022:123389. https://doi.org/10.1016/j.jorganchem.2024.123389
Ocampo AYD, Salazar RE, Ramírez CMÁ, Rios MY (2023) A comprehensive review of liquid chromatography methods for fumonisin determination, a 2006–2022 update. Arab J Chem 16:104716
OECD (2007) Report on the test results of endocrine disrupting effects of tributyltin [TBT] on fish [Draft]. http://www.oecd.org/dataoecd/7/1/2461193.pdf
Pallavi MS, Harischandra NR, Pavankumar K, Ratnamma N, Shwetha A, Naveenkumar P, Paramasivam M, Udaykumar NR, Prabhuraj A, Bheemanna M (2023) Determination of 73 multi-class pesticides in okra (Abelmoschus esculentus L.) fruits using LC-MS/MS and GC-MS/MS and estimation of analytical uncertainty of measurement. Food Chem 19:100814
Pathan IR, Patel MK (2023) A comprehensive review on the synthesis and applications of Schiff base ligand and metal complexes: a comparative study of conventional heating, microwave heating, and sonochemical methods. Inorg Chem Commun 158:111464
Qadeer A, Rui G, Ya QL, Ran D, Liu CY, Jing D, Anis M, Liu MY, Wang SH, Jiang X et al (2023) A mega study of antibiotics contamination in eastern aquatic ecosystems of China: occurrence, interphase transfer processes, ecotoxicological risks, and source modeling. J Hazard Mater 458:131980
Retnam CTG, Rose SV, Kumari BS (2023) Synthesis, characterization, biological activity and molecular docking study of transition metal complexes from heterocyclic ligand system. J Mol Struct 1282:135162
Soldevila-Barreda JJ, Azmanova M, Pitto-Barry A, Cooper PA, Shnyder SD, Barry NPE (2020) Preclinical anticancer activity of an electron-deficient Organoruthenium(II) complex. ChemMedChem 15:982–987
Souza HDO, Costa RDS, Quadra GR, Fernandez MADS (2021) Pharmaceutical pollution and sustainable development goals: going the right way? Sustain Chem Pharm 21:100428
Temerdashev Z, Abakumov P, Bolshov M, Abakumova D, Pupyshev A (2023) Possibilities and limitations of ICP-spectrometric determination of the total content of tin and its inorganic and organic speciations in waters with different salinity levels—part 2: separate determination of inorganic and organic speciations of tin. Molecules 28:6615. https://doi.org/10.3390/molecules28186615
Turner APF (2013) Biosensors: sense and sensibility. Chem Soc Rev 42(8):3184–3196
Vigneshvar S, Sudhakumari CC, Senthilkumaran B, Prakash H (2016) Recent advances in biosensor technology for potential applications–an overview. Front Bioeng Biotechnol 4:11
Wang J, Chen G, Jiang H, Li Z, Wang X (2013) Advances in nano-scaled biosensors for biomedical applications. Analyst 138(16):4427–4435
Wang B, Takahashi S, Du X, Anzai JI (2014) Electrochemical biosensors based on ferroceneboronic acid and its derivatives: a review. Biosensors 4(3):243–256
Wang SQ, Chen HT, Sun BG (2020) Recent progress in food flavor analysis using gas chromatography-ion mobility spectrometry (GC-IMS). Food Chem 315:126158
Wang DY, Liu R, Guo W, Li G, Fu Y (2021) Recent advances of organometallic complexes for rechargeable batteries. Coord Chem Rev 429:213650
Waters JE, Stevens-Cullinane L, Siebenmann L, Hess J (2023) Recent advances in the development of metal complexes as antibacterial agents with metal-specific modes of action. Curr Opin Microbiol 75:102347. https://doi.org/10.1016/j.mib.2023.102347
Wei YL, Wang LP, Liu J (2023) The diabetogenic effects of pesticides: evidence based on epidemiological and toxicological studies. Environ Pollut 331:121927
Yadav S, Jijoe SP, Thinley T, Hosakote A, Anil KKM, Shivaraju HP (2023a) Hybrid $ZnFe_2O_4/Ag_2S$ nanocomposite for enhanced photocatalytic and bacterial activity towards targeted superbugs. Mater Today Proc 75:46–57

Yadav N, Maurya BM, Chettri D, Pooja, Pulwani C, Jajula M, Kanda SS, Babu HWS, Elangovan A, Velusamy P et al (2023b) Artificial intelligence in heavy metals detection: methodological and ethical challenges. Hyg Environ Health Adv 7:100071

Yahya M, Yaman M, Seferoğlu Z (2022) Optical chemosensors: principles, chemistry, strategies, and applications. IntechOpen, London. https://doi.org/10.5772/intechopen.105968

Yang QQ, Li ZY, Lu XN, Duan QN, Huang L, Bi J (2018) A review of soil heavy metal pollution from industrial and agricultural regions in China: pollution and risk assessment. Sci Total Environ 642:690–700

Yang Q, Zhao Y, Yu H, Xiong X, Huang K (2023) Recent advances of sample separation and preconcentration techniques for analytical atomic spectrometry. Appl Spectrosc Rev. https://doi.org/10.1080/05704928.2023.2231537

Younis AM (2020) Accumulation and rate of degradation of organotin compounds in coastal sediments along the Red Sea, Egypt. Egypt J Aquat Biol Fish 24(5):413–436. https://doi.org/10.21608/ejabf.2020.108918

Zhang J, Pitto-Barry A, Shang L, Barry NPE (2017) Anti-inflammatory activity of electron-deficient organometallics. R Soc Open Sci 4:170786. https://doi.org/10.1098/rsos.170786

Zhang S, Wang B, Li S, Xiao HL, Liu G, Zhang Z, Wang X (2024) Metal-/carboxylate-directed four D 10 piperazine-amide-based coordination polymers for the fluorescent detection of nitrophenol and nitroaniline in various water environments. J Mol Struct 1297:136929

Zhu GS, Jia Y, Ding JY, Yin H, Chen Y, Yu BY, Zheng YY, Verpoort F (2024) Design and construction of a mixed-ligand coordinated fluorescent complex and its application for sensing ions, antibiotics, and pesticides in aqueous solution. Inorganics 12:93. https://doi.org/10.3390/inorganics12040093

Ziolkowska A, Doroszuk J, Ponikiewski L (2023) Overview of the synthesis and catalytic reactivity of transition metal complexes based on C=P bond systems. Organometallics 42:505–537

The manufacturer's authorised representative in the EU is Springer Nature Customer Service Centre GmbH, Europaplatz 3, 69115 Heidelberg, Germany. If you have any concerns regarding our products, please contact ProductSafety@springernature.com

Printed and bound by CPI Group (UK) Ltd, Croydon, CR0 4YY
01/10/2026
02225883-0015